Mesozoic and Cenozoic paleocontinental maps

A. G. SMITH
Lecturer in Geology
University of Cambridge

J. C. BRIDEN
Professor of Geophysics
University of Leeds

CONTENTS

CAMBRIDGE UNIVERSITY PRESS
CAMBRIDGE
LONDON · NEW YORK · MELBOURNE

CAMBRIDGE UNIVERSITY PRESS
Cambridge, New York, Melbourne, Madrid, Cape Town,
Singapore, São Paulo, Delhi, Tokyo, Mexico City

Cambridge University Press
The Edinburgh Building, Cambridge CB2 8RU, UK

Published in the United States of America by Cambridge University Press, New York

www.cambridge.org
Information on this title: www.cambridge.org/9780521291170

© Cambridge University Press 1977

This publication is in copyright. Subject to statutory exception
and to the provisions of relevant collective licensing agreements,
no reproduction of any part may take place without the written
permission of Cambridge University Press.

First published 1977
Reprinted 1979
Re-issued 2011

A catalogue record for this publication is available from the British Library

Library of Congress Cataloguing in Publication data
Smith, Alan Gilbert, 1937–
Mesozoic and Cenozoic paleocontinental maps.
(Cambridge earth science series)
Bibliography: p.
1. Plate tectonics – Maps. 2. Continents – Maps, outline and base. I. Briden, J. C., joint author. II. Title.
G1046.C55S6 1976 912'.1'55141 76–14052

ISBN 978-0-521-29117-0 Paperback

Cambridge University Press has no responsibility for the persistence or
accuracy of URLs for external or third-party internet websites referred to in
this publication, and does not guarantee that any content on such websites is,
or will remain, accurate or appropriate.

Introduction

We hope that the maps in this book will be useful to all students, teachers and research workers concerned with large-scale geological problems. There are fifty-two maps in all, drawn in four series at thirteen periods of time: present-day, 10 Ma, 20 Ma, and at 20-Ma intervals before that back to 220 Ma. The four series consist of a Mercator series, two polar stereographic series and a Lambert equal-area series. The maps may be regarded as modifications and interpolations of the 220 Ma, 170 Ma, 100 Ma and 50 Ma paleocontinental maps published previously (Smith, Briden & Drewry, 1973; Briden, Drewry & Smith, 1974).

As the title shows, the emphasis is on the past relative positions of the continents. Of necessity the oceans separate the continents from one another, but no attempt has been made to show any former oceanic feature, other than the approximate edges of the ocean basins. It is hoped that readers will plot their own paleogeographic, paleontologic or paleoclimatic data on these maps, or use them to make their own plate tectonic interpretations of the time concerned.

Most of us depend heavily on the recognition of coastlines to locate our positions on present-day geographic maps. Yet the coastline is one of the most ephemeral features of paleogeography and only in exceptional situations can it be located in the stratigraphic record. Thus the coastlines drawn on the maps in this book are merely our estimates of the past positions of present-day coastlines. Since these features did not exist in the past, their main value is that of aiding the recognition of the past positions of continental fragments. The continental shelves at the edges of the continents are much more enduring than the coastlines. The maps show the present-day 1000-metre (500-fathom) submarine contour, except around most of the Pacific basin, where it has been omitted.

The past positions of the present geographic latitude–longitude grid are shown within the continental fragments at intervals of ten degrees so that past features on the continents may be plotted in their past, original positions. The problems of drawing the present grid in areas affected by orogeny, and of drawing former continental boundaries of areas that have subsequently been welded together, are briefly discussed below (pp. 5 and 8–9).

Superposed on each map is a paleogeographic latitude–longitude grid drawn at thirty-degree intervals. Nevertheless, the maps cannot be regarded as paleogeographic maps since they do not show past geography. They do show our estimates of former relative continental positions. In other words, they are paleocontinental maps.

Apart from the fact that global paleocontinental maps have not been published previously for most of the time intervals shown here, an interesting feature of the figures is that they are machine made. Only captions have been added to the computer drawings. Though this method of production gives maps that may not be as pleasing to look at as those drawn by hand, it has two distinct advantages: the maps are available quickly and they cost less to make. It also means that subsequent editions can be readily corrected when new data make it necessary to do so. It cannot be overemphasized that these maps are provisional estimates based on our interpretation of the published data (p. 8).

Method of making the maps

The maps are made in two stages. The first stage is essentially the making of a continental reassembly after closure of the Atlantic and Indian Oceans by the appropriate amount. The second stage projects the resultant paleocontinental reassembly as a map. The method is more fully described elsewhere (Smith *et al.*, 1973), and a short summary only is given below.

Motion between two continents takes place at one or more plate boundaries. At a given instant the motion along a particular plate boundary may be described as a rotation at a given rate about an axis passing through the Earth's centre. The motion throughout a time interval may be found by summing all the instantaneous motions in that interval. The sum is a finite rotation about an axis passing through the Earth's centre. The net motion between two continents separated by more than one active plate boundary in a time interval is simply the sum of the finite rotations that have taken place across each of the plate boundaries in that time. Because finite rotations are not commutative, that is, they do not add like vectors, care must be taken in the order in which the finite rotations are added together.

The finite rotations taking place across compressional plate margins (present-day subduction zones) cannot be inferred from the margins themselves or from the effects adjacent to the margins. Thus in the absence of any other information, two continents whose relative motion involves the action of a compressional plate margin in the time interval concerned cannot be repositioned relative to each other.

In principle, the finite rotations taking place across extensional and/or translational plate margins can be determined. In general such motions create aseismic ocean basins like most of the present-day Atlantic or Indian Oceans. Providing adequate surveys have been made of such areas, the finite rotations necessary to describe the relative motions of the surrounding continents can readily be estimated. In most cases the rotations are obtainable by matching the corresponding pairs of ocean-floor magnetic anomalies. Earlier shapes and sizes of the presently expanding ocean basins may be estimated by 'winding back' the ocean floor by the amount the basins have grown since the time of interest.

Because most of the floor that has ever been formed within the Atlantic and Indian Oceans is still preserved in those oceans, the former relative positions of the continents around those oceans may be estimated at all times since their creation. To make a continental reassembly, one of the continents is chosen as a reference and all the others are repositioned relative to it. Africa has been used as the reference continent for the present series of maps, but any other continent or continental fragment around the Atlantic or Indian Oceans could equally well have been used as reference.

Available data are sufficiently widely distributed (Pitman, Larson & Herron, 1974), that all the major continental fragments may be approximately repositioned relative to one another as far back in time as the beginning of opening of the Atlantic and Indian Oceans. The oldest known part is the Atlantic Ocean that lies between Africa and North America. It probably began to form in early Jurassic time. Some parts of the Indian Ocean may be of a similar age but the data are inadequate to show whether this is so. The finite rotations required to make the reassemblies have been taken from the references given by Pitman *et al.* (1974) to individual marine geophysical surveys.

The three series of maps at 180, 200 and 220 Ma show one supercontinent – Wegener's Pangaea. The only difference between these maps is in the orientation of Pangaea relative to the paleogeographic grid.

By choosing the appropriate route it is possible to circumvent the problem of repositioning continents separated by compressive plate boundaries at some time in the past 220 Ma. For example, despite the growth of the Alpine chains, Africa may be repositioned relative to Europe by repositioning Africa relative to North America, and then North America relative to Europe. Pangaea is believed to have been created by the coalescence of at least three large continents in later Paleozoic time. The positions of these continents relative to one another cannot yet be determined because all routes in the repositioning procedure run across compressional plate boundaries. Some constraints that may be placed on their relative positions are discussed in Smith *et al.* (1973).

The second stage of map-making consists in estimating the position of the paleogeographic poles on the reassembly and making a map projection. The best estimate of the past positions of the geographic poles appears to be the mean positions of the paleomagnetic poles. Most of the paleomagnetic data used in these maps have been published in the compilations of Irving and of McElhinny. These are referenced in McElhinny (1972, 1973). Those poles satisfying certain reliability criteria (McElhinny, 1973) have been selected for making the maps.

To make a map, all the reliable north paleomagnetic poles lying within the stable parts of all the continental fragments that can be repositioned relative to one another have been examined. Those north poles whose age range lies within 10 Ma of the age of the reassembly are selected for the map. They are all rotated to the reference continent using the rotations that have been employed to make the reassembly. The mean paleomagnetic north pole of the rotated north poles is calculated. This mean pole is taken as the best estimate of the north geographic pole of the reassembly relative to the reference continent. Once the pole position is known, the paleogeographic latitude–longitude grid may be superimposed on the reassembly. The zero meridian of longitude is arbitrary, as in present-day maps. But all longitude differences among areas whose relative positions can be determined on the reassembly are fixed by their positions relative to the mean pole of the reassembly.

All of these operations are carried out automatically by suitable computer programs. The end result is a set of punched cards that will produce a paleocontinental map when read in conjunction with a digitized world map broken up into suitable continental fragments. The map-making program (a modified version of R. L. Parker's SUPERMAP program) will also draw maps on projections other than those presented here.

Projections

The projections used are Mercator's, stereographic and Lambert's equal-area projections. The stereographic projections are in two series: a north polar and a south polar series. Lambert's equal-area series consists of an equatorial view only. All four series serve different needs. Ellipticity corrections are negligible for world maps on the scale used.

Mercator's projection

Mercator's projection is obtained by projecting geographical outlines onto a cylinder tangential to the equator. The latitude lines are thus of equal length and are undistorted only at the equator. The separation of the latitude lines is adjusted so that the scale along the longitude lines at any point is the same as the scale of the latitude lines at the same point. The higher latitude lines are invariably stretched in length, but the projection gives an equal stretching of the longitude lines to preserve the equality of scale in all directions at each point.

Because of this, the projection is said to be orthomorphic. Small areas are projected without changes of shape; only the relative sizes of the projected areas alter. The stretching of the latitude lines causes areas nearer the poles to have larger projected areas than do areas nearer the equator. This is the cause of the familiar enlarged size, but locally correct shape, of Greenland on Mercator maps.

The principal advantages of Mercator's projection for global geological problems are its familiarity, its high-lighting of the tropical areas, and the parallelism of latitude and longitude lines on the map. The main disadvantages of the normal Mercator, as opposed to the transverse Mercator projection, are the inability to represent the polar areas (the poles themselves are at infinity) and the areal exaggeration of areas near the poles. The maps shown extend to seventy degrees of latitude north and south (Maps 1–13).

The construction is mathematically simple. The latitude line at the equator has a length $2\pi r$, where r is the radius of the globe and of the tangential cylinder. The spacing between longitude lines, or meridians, is therefore $2\pi r\,(\Phi_1 - \Phi_2)/360$, where Φ_1 and Φ_2 are the values of the longitudes in degrees. The distance from a particular latitude line to the equator is $r \cdot \log_e(\tan 45° + \lambda/2)$, where λ is the latitude in degrees. Clearly if $\lambda = 90°$, the line is at infinity, since $\tan[45 + (90/2)]$ is $\tan 90°$, which is infinite, and its $\log_e$ is also infinite.

Stereographic projection

The stereographic projection is obtained by projecting the geographic outlines from a point on the surface of the sphere onto a plane normal to the diameter through the projection point. The polar stereographic net consists of latitude lines that are circles centred on the projection poles and longitude lines that are straight and pass through the pole, like the spokes of a wheel. An advantage of the projection is its familiarity to many geologists.

The projection resembles Mercator's in that the scale along the meridians at a point is the same as the scale along the latitude lines. Thus it, too, is orthomorphic. Small shapes are preserved unchanged by the projection, but the area of a given shape changes from place to place. Near the pole of the projection the size of a shape is smaller than it is towards the margins.

Estimates of angular relationships, of distances between two points, and so on, may be made using standard stereographic nets. Such estimates may be particularly useful for plate tectonic problems. The advantage of the polar projections used here (Maps 14–39) is the highlighting of those areas that cannot be shown on a normal Mercator projection. Its disadvantages are similar to Mercator's: it is impossible to show the whole globe on one projection and there are considerable exaggerations in scale at the edges compared with the centre.

Mathematically the construction is simple. The meridians are straight lines passing through the projection pole. For two longitude lines at Φ_1 and Φ_2, the angle of intersection is simply $(\Phi_1 - \Phi_2)$. If r is the radius of the globe that is being projected and the projection plane passes through the centre of the sphere, it is easy to show that the radius of the latitude circle representing the latitude λ is $r \cdot \tan(45 - \lambda/2)$. Thus the radius of the equator is r, or the radius of the original globe. When $\lambda = -90°$, the value of the latitude of the point opposite the projection pole, the radius is $2r \cdot \tan[45 - (-90/2)]$, i.e. infinite.

Lambert's equal-area projection

As its name indicates, this projection preserves relative differences among the areas on the globe. Its construction is more complex than that of the two previous projections, and a full description is not given here.

In the series shown (Maps 40–52), the centre of the projection is 'zero' longitude on the equator. The best way to see the relationships among the various geographic features is to examine the present-day world (Map 40).

The positions of the polar regions are obvious from the intersection of the present-day geographic grid. Encircling the poles are ovals of constant latitude drawn at thirty-degree intervals. The front hemisphere is bounded by the circle that cuts through both poles. Beyond it on the map or behind it on the globe is the 'back hemisphere'. Inspection shows that there are some very peculiar spatial relationships among all points at the edges of the map. These are the points that lie close to the point on the equator directly opposite the projection centre. The relationships among the points are so confusing that, for interest, they are shown only on the present-day map. On all other maps, points forming about ten per cent of the area of the back hemisphere closest to the back projection centre have been omitted by masking. Another effect of the projection is that the subtropical regions form two doughnut-shaped regions joined at the equator. Their shape is best seen by subtracting all regions polewards of the thirty-degree latitude line. Reference to the present-day map is the easiest way of seeing the geometry of the projection.

The principal advantages of the projection are its ability to show the two polar regions on one map and also its equal-area property. This is particularly useful for work involving global distributions of sediment types, faunal provinces, etc. Its disadvantage is its inability to show geographical relationships among the tropical regions, though this could easily be done by changing the position of the projection centre.

Reliability of the maps

There are two sources of error. The first lies in the construction of the reassembly and the second in projection of this reassembly as a map. The first source includes several different kinds of errors. There are errors due to the uncertainties in the ocean-floor-spreading histories of the Atlantic and Indian Oceans; those caused by ignorance of the past positions of all continental fragments affected by orogeny; and those attributable to a lack of knowledge of the shapes and former boundaries of continental fragments that have collided with one another.

The relative positions of the continents are best known for the past 80 Ma. Prior to this time the ocean-floor-spreading anomalies are less frequently developed, less well dated and in some cases they may have been eliminated, possibly by submarine diagenetic processes. These problems are particularly acute for the earlier history of the Indian Ocean, where the data are also poorly distributed.

In two previous publications the South Atlantic and Indian Oceans were assumed to have begun to open at 100 Ma (Smith *et al.*, 1973; Briden *et al.*, 1974). Subsequent work suggests, but does not prove, that the South Atlantic began to open much earlier at 140 Ma. If this is so, then the Indian Ocean between Africa and Antarctica probably began to open at the same time. We have presented this possibility in this book. In our opinion, the present maps are more likely to be more correct, but there is no firm evidence for or against this view. Perhaps this situation brings out clearly the provisional nature of the maps.

Similar problems exist in the North Atlantic region, where the successive positions of Greenland have been estimated from available data, but may well have differed from those shown. The initial positions of the continents around the Atlantic Ocean are taken from Bullard, Everett & Smith (1965).

The absence of adequate survey data for the earliest stages of the opening of parts of the Indian Ocean also introduces uncertainties into the fit of the southern continents. Thus it is not yet known whether the Gondwanaland reassembly of the 180 Ma map is correct (Smith & Hallam, 1970). The relative positions of South America, Africa and Arabia, and of Australia and Antarctica are well known, but how South America–Africa–Arabia and Australia–Antarctica join up with India, Ceylon, and Madagascar is not yet settled, even on the scale of the maps.

The positions of all those areas affected by Mesozoic and Tertiary orogenesis is unknown. The relative initial positions of the continental fragments around the Caribbean and the Mediterranean and their evolution in time are based mostly on speculations by Freeland & Dietz (1971) and Smith (1971) respectively. The Caribbean area has been held fixed to North America. On maps of 160 Ma and older it overlaps onto Africa. No attempt has been made to correct this overlap. Though relative motions are known to have occurred among the continental fragments bordering the Pacific basin, no attempt has been made to reconstruct these areas. The approximated outcrop areas of the Mesozoic and Tertiary orogenic belts are shown in the Mercator series of maps in Smith *et al.* (1973), or can be found in other publications.

Prior to the collision of two or more continental fragments it may be assumed that an oceanic region existed between them. After collision the former boundaries merge into a single continental region. The line (or lines) of joining together have been arbitrarily estimated on the maps and are shown by lines of crosses.

The second source of uncertainty lies in the estimate of the mean paleomagnetic pole of the continental reassembly. The number N in the caption of each map refers to the number of separate paleomagnetic studies that have been used to make the map. A study has been accepted provided it satisfies certain criteria (McElhinny, 1973) and provided too that it has an age range lying within 10 Ma of the time for which the map is required. For example, the 40 Ma map draws on all reliable pole studies on the stable parts of the continents whose age ranges include some part of the interval 30–50 Ma. Poles that lie within deformed regions have been excluded. The age criterion for accepting poles has some disadvantages. In particular, it means that a pole with a poorly determined age range will appear in the pole list for a much larger number of maps than one with a precisely determined age range. It would be better to weight poles according to the precision of their determined ages, but we have not done this.

Alpha-95 in the caption of each map is a standard statistical measure for the spread of data on a sphere. Essentially it is the radius in degrees on the surface of the sphere, of a circle centred on the mean of the data. The ninety-five per cent level means that there is a one in twenty chance that the true mean lies outside the circular limit. All of the confidence limit circles have radii smaller than ten degrees, and some are smaller than five degrees. This is a small dispersion for paleomagnetic data, though the estimated dispersion may in some cases be misleadingly small. This situation arises in cases where most of the poles come from only one or two continental fragments, rather than being uniformly distributed among all the fragments. Because each pole study has been treated as a completely independent set of data – whether or not it has been made on one continent – the scatter of the poles may appear much tighter than it might were the poles to come from several continents. Nevertheless, the scatter of poles is remarkably small, and gives some idea of the uncertainties in the orientation of the reassembly.

Further unavoidable weighting of the data arises from the variation of magnetic inclination I with the latitude λ of a dipole field:

$$\tan I = 2 \tan \lambda.$$

An error $\mathrm{d}I$ in inclination gives rise to an error $\mathrm{d}\lambda$ in the location of the pole on the paleomeridian:

$$\mathrm{d}\lambda = \tfrac{1}{2}(1+3\sin^2\lambda)\,\mathrm{d}I,$$

and this uncertainty increases with paleolatitude.

Another source of error that affects the construction both of the reassembly and of the map is the interrelationship between the fossil, magnetic reversal and isotopic time scales. With a few exceptions, the name given to a map has been obtained from the stratigraphic scale of Harland *et al.* (1967). The age in Ma has been obtained by using this scale in conjunction with that of Harland, Smith & Wilcock (1964). The main exceptions to this procedure are that the name Paleocene has been used instead of Danian, and that no stratigraphic name has been given to the early Triassic map.

The ages in Ma of the ocean-floor magnetic anomaly time scale have been obtained from Heirtzler *et al.* (1968). The ages in Ma of paleomagnetic poles have been taken as their isotopic age (where known), or calculated by converting the stratigraphic range in the pole compilations of Irving and of McElhinny into Ma, using the time scale of Harland *et al.* (1964). No amendments to the above scales have been incorporated into the data.

The total effect of all the above sources of error is not known and is difficult to estimate.

Final remarks

The maps are based on quantitative geophysical or topographic data. Lack of space prohibits the presentation of most of the information used to make them, but it can be obtained from the references in the text. Lack of space also prevents the inclusion of a series of Lambert equal-area 'back view' maps such as appear in Briden *et al.* (1974). These are particularly useful for areal measurements in the Pacific region. They are available as a set of thirteen maps at about the same scale as those in the book, price £1.50, from A. G. Smith.

Acknowledgements

The maps are the outcome of projects supported by the Natural Environment Research Council (GR/3/1287 and GR/3/2277). Mrs G. E. Drewry and Mrs A. M. Hurley, research assistants for this work, ran all the computer programs that made the maps. Their help is gratefully acknowledged. The authors also thank Mr W. B. Harland and Mr N. F. Hughes for their encouragement and for suggesting improvements to the text.

Bibliography

Briden, J. C., Drewry, G. E. & Smith, A. G. (1974). Phanerozoic equal-area world maps. *J. Geol.*, **82**, 555–74.

Bullard, E. C., Everett, J. E. & Smith, A. G. (1965). The fit of the continents around the Atlantic. In: A symposium on continental drift. *Phil. Trans. R. Soc., London, Ser. A*, **258**, 41–51.

Freeland, G. L. & Dietz, R. S. (1971). Plate tectonic evolution of Caribbean and Gulf of Mexico Region. *Nature, Lond.*, **232**, 20–23.

Harland, W. B., Smith, A. G. & Wilcock, B. (1964). *The Phanerozoic Time-scale.* (Symposium dedicated to Professor Arthur Holmes.) *Q. J. Geol. Soc., London*, **120S**, 1–458.

Harland, W. B. *et al.* (ed.) (1967). *The Fossil Record.* London: Geological Society. 828 pp.

Heirtzler, J. R., Dickson, G. O., Herron, E. M., Pitman, W. C. & Le Pichon, X. (1968). Marine magnetic anomalies, geomagnetic field reversals and motions of the ocean floor and continents. *J. Geophys. Res.*, **73**, 2119–36.

McElhinny, M. W. (1972). Notes on progress in geophysics, palaeomagnetic directions and pole positions. XIII. Pole numbers 13/1 to 13/94. *Geophys. J. R. Astron. Soc.*, **30**, 281–93.

McElhinny, M. W. (1973). *Palaeomagnetism and Plate Tectonics.* London: Cambridge University Press. 358 pp.

Pitman, W. C., Larson, R. L. & Herron, E. M. (1974). *The Age of the Ocean Basins.* Boulder, Colorado: Geological Society of America.

Smith, A. G. (1971). Alpine deformation and the oceanic areas of the Tethys, Mediterranean and Atlantic. *Bull. Geol. Soc. America*, **82**, 2039–70.

Smith, A. G., Briden, J. C. & Drewry, G. E. (1973). Phanerozoic world maps. In: Organisms and continents through time, ed. N. F. Hughes. *Palaeontology, Special Papers*, **12**, 1–42.

Smith, A. G. & Hallam, A. (1970). The fit of the southern continents. *Nature, Lond.*, **225**, 139–44.

Map 1
Present day

Mercator

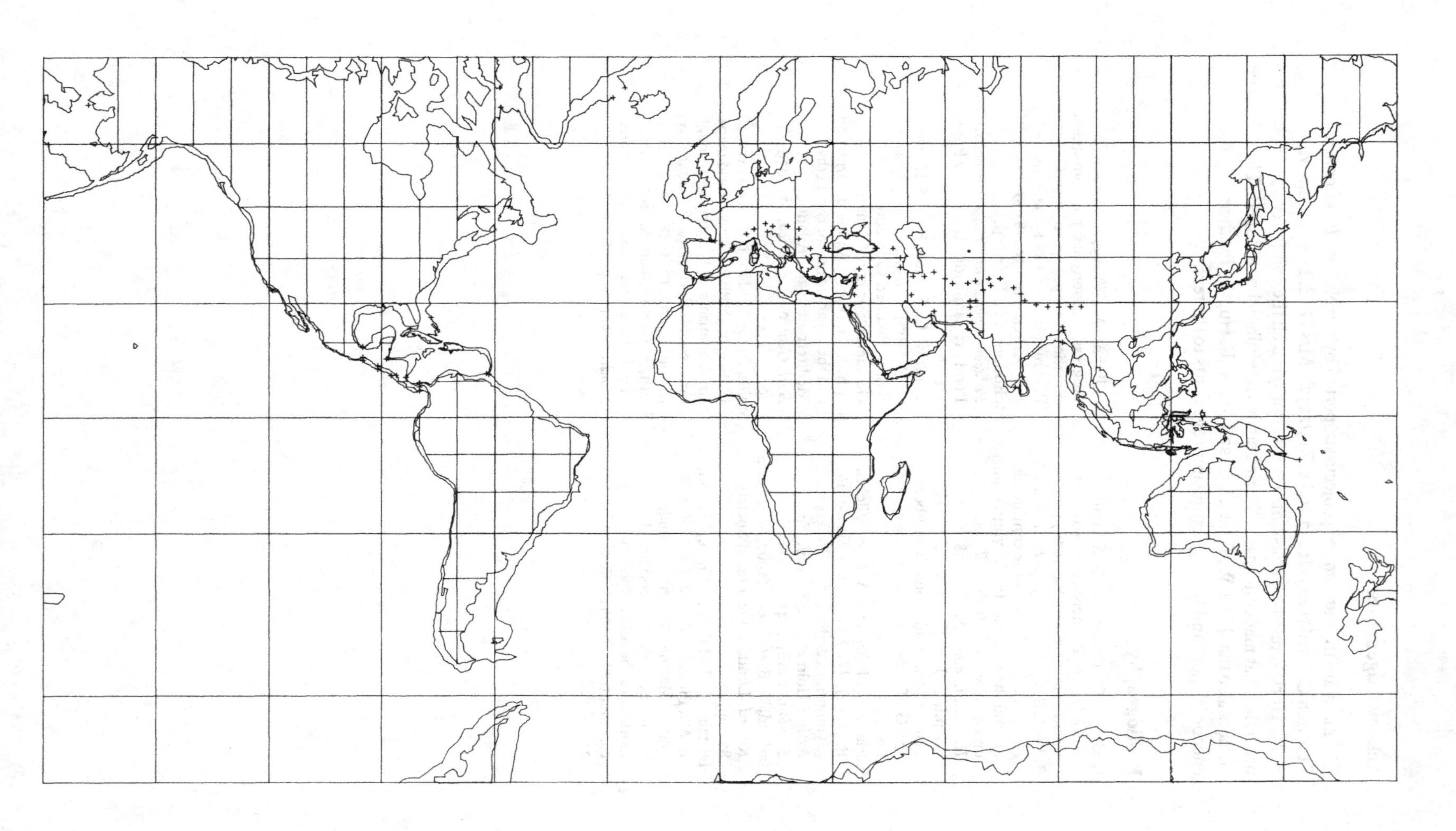

Map 2

10 million years
late Miocene (Cenozoic)

Mercator
$N = 87$ Alpha-95 = 3·2

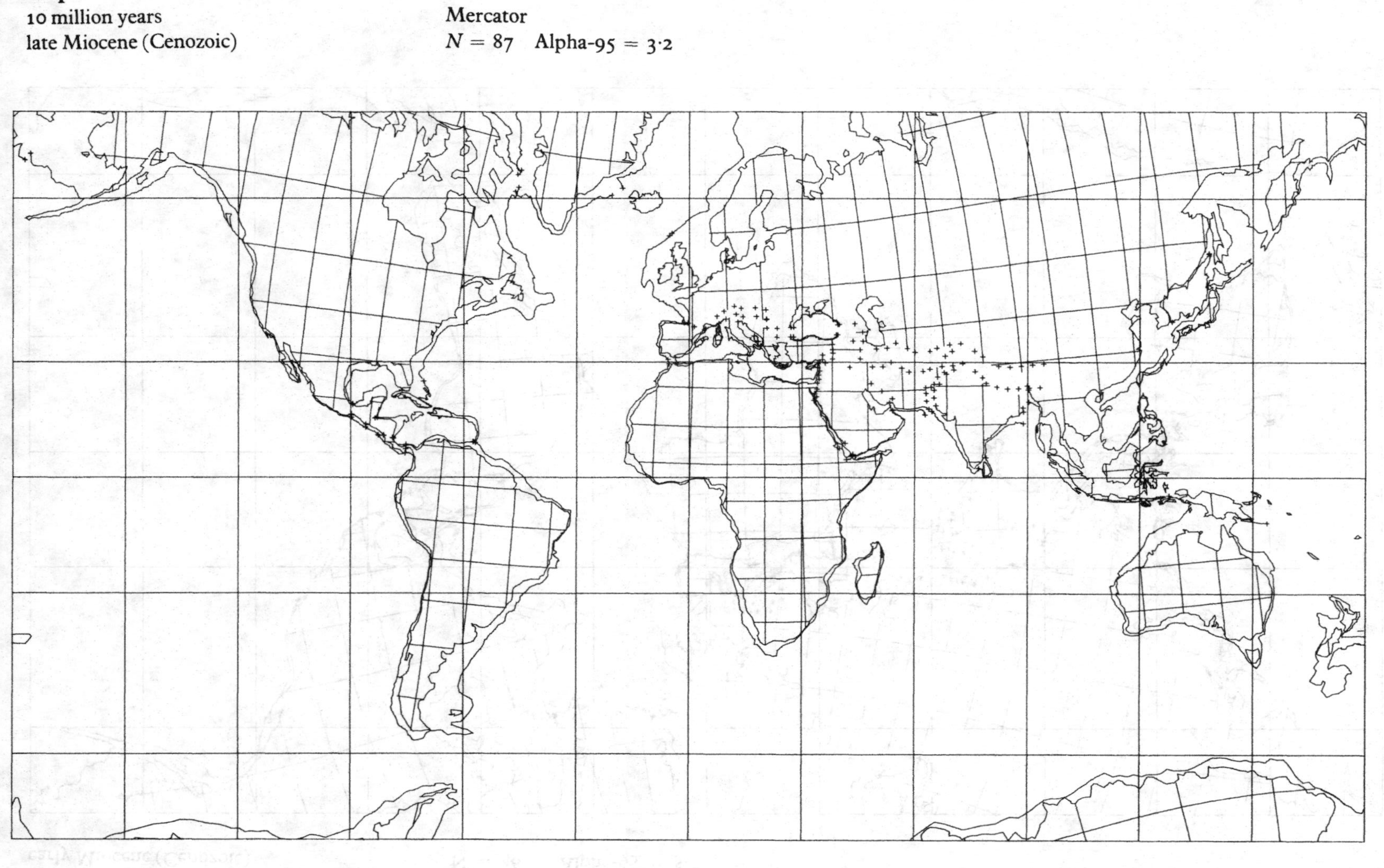

Map 3

20 million years

early Miocene (Cenozoic)

Mercator

N = 46 Alpha-95 = 5·7

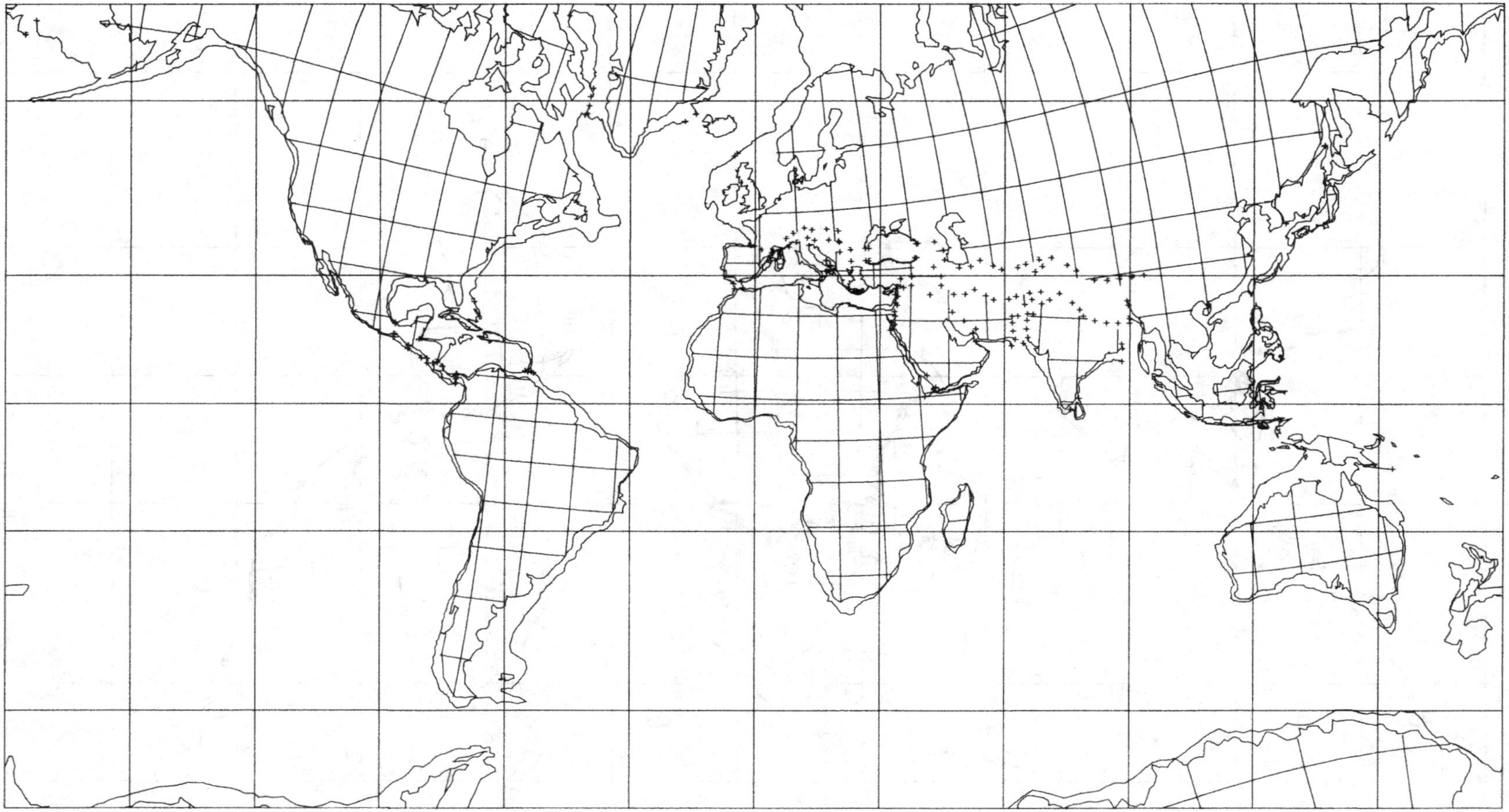

Map 4
40 million years
late Eocene (Cenozoic)

Mercator
$N = 28$ Alpha-95 = 4·5

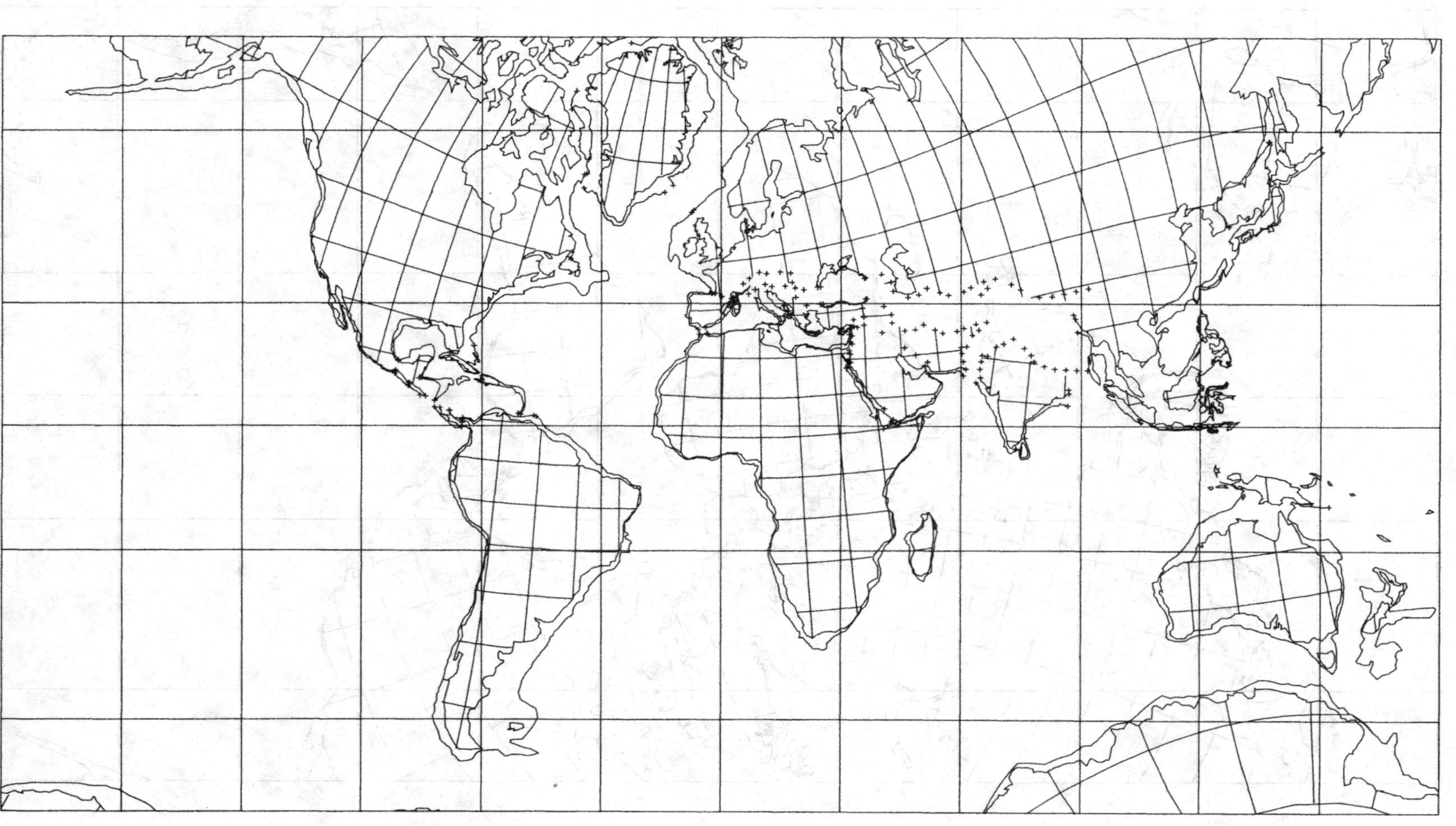

Map 5
60 million years
Paleocene (Cenozoic)

Mercator
$N = 43$ Alpha-95 = 4·7

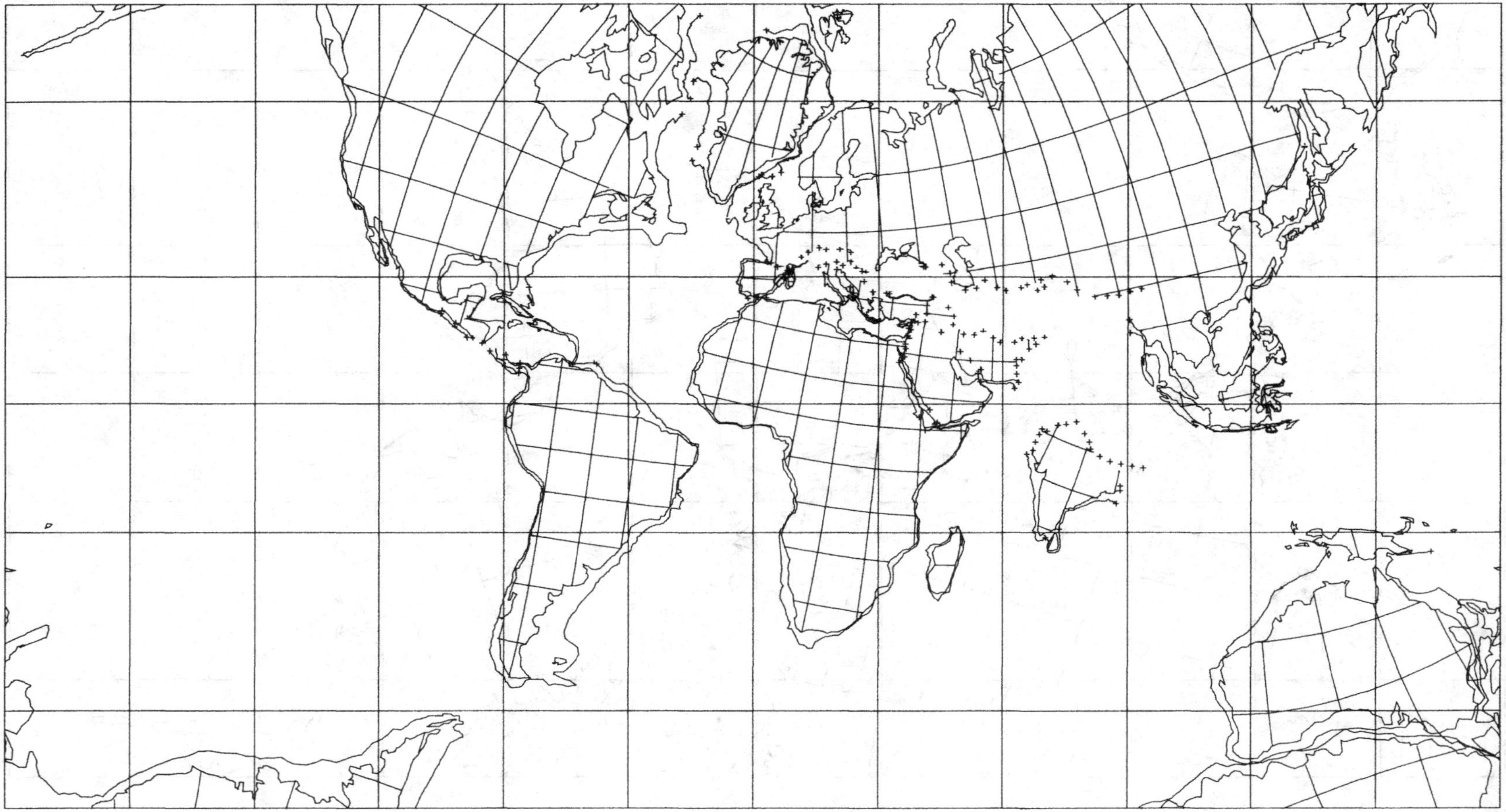

Map 6

80 million years

Santonian (late Cretaceous)

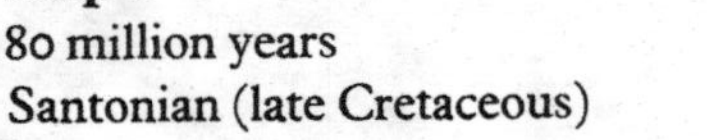

Mercator

$N = 25$ Alpha-95 = 6·0

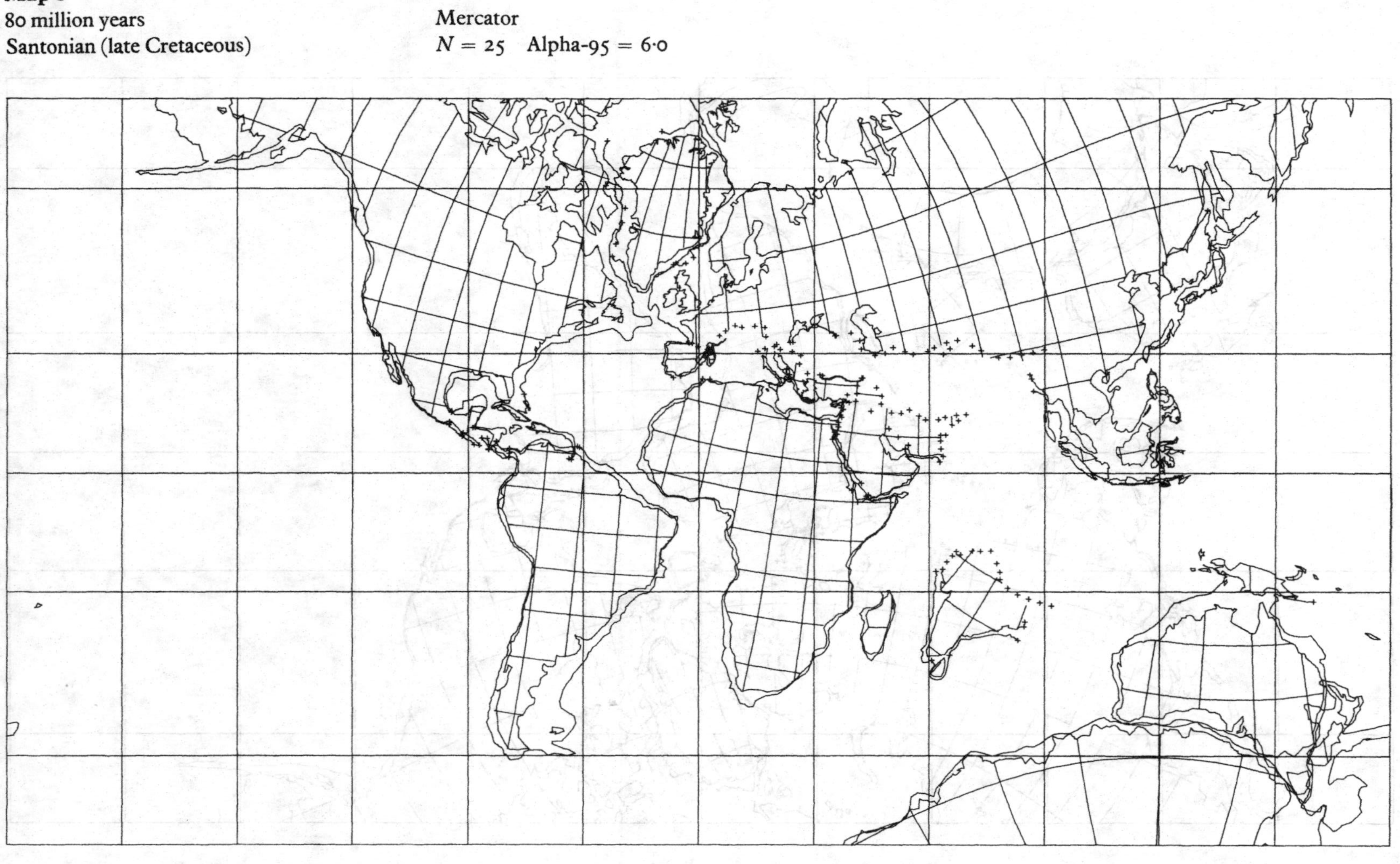

Map 7
100 million years
earliest Cenomanian (mid-Cretaceous)

Mercator
$N = 43$ Alpha-95 = 5·2

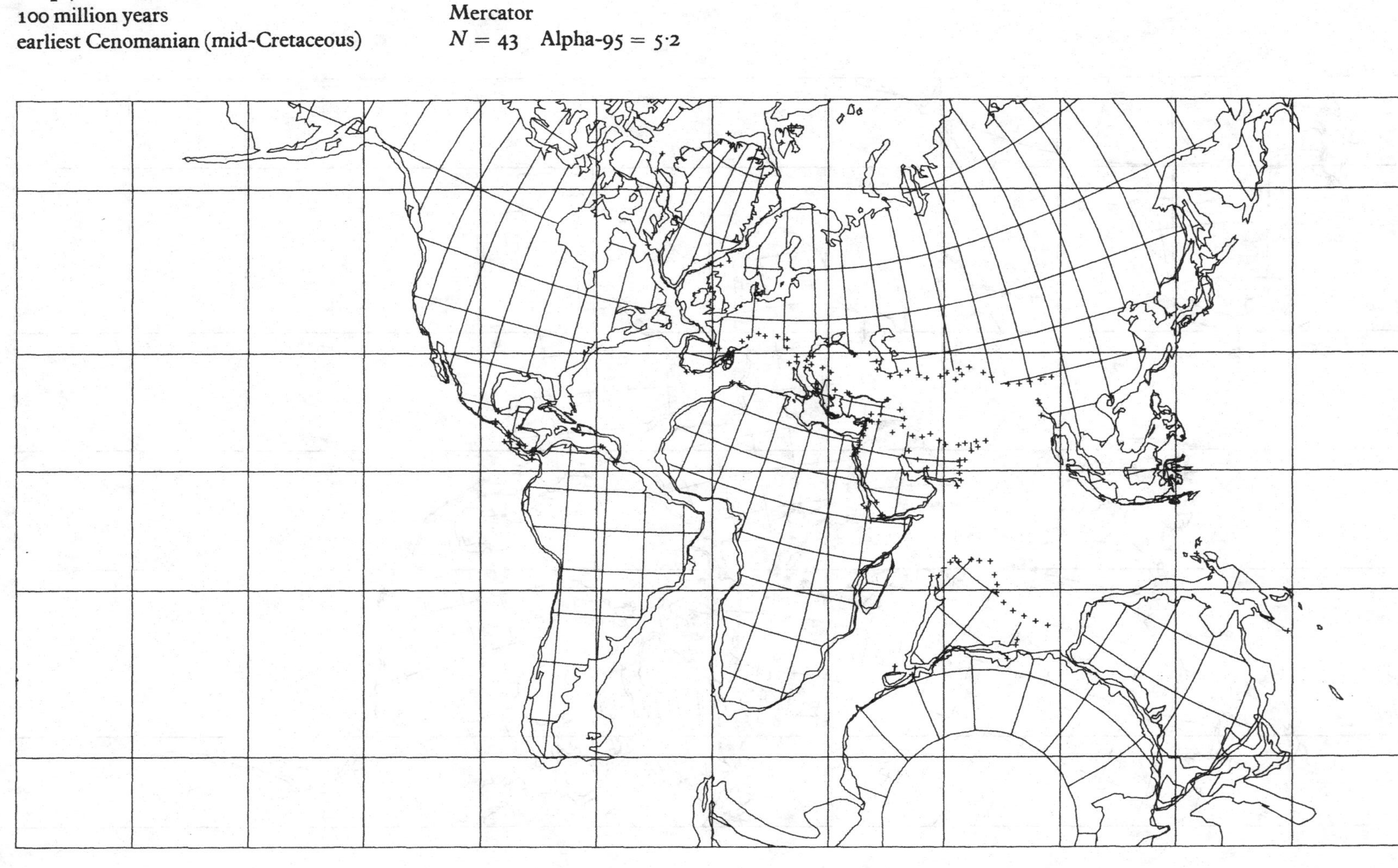

Map 8

120 million years

Hauterivian (early Cretaceous)

Mercator

$N = 27$ Alpha-95 = 6·2

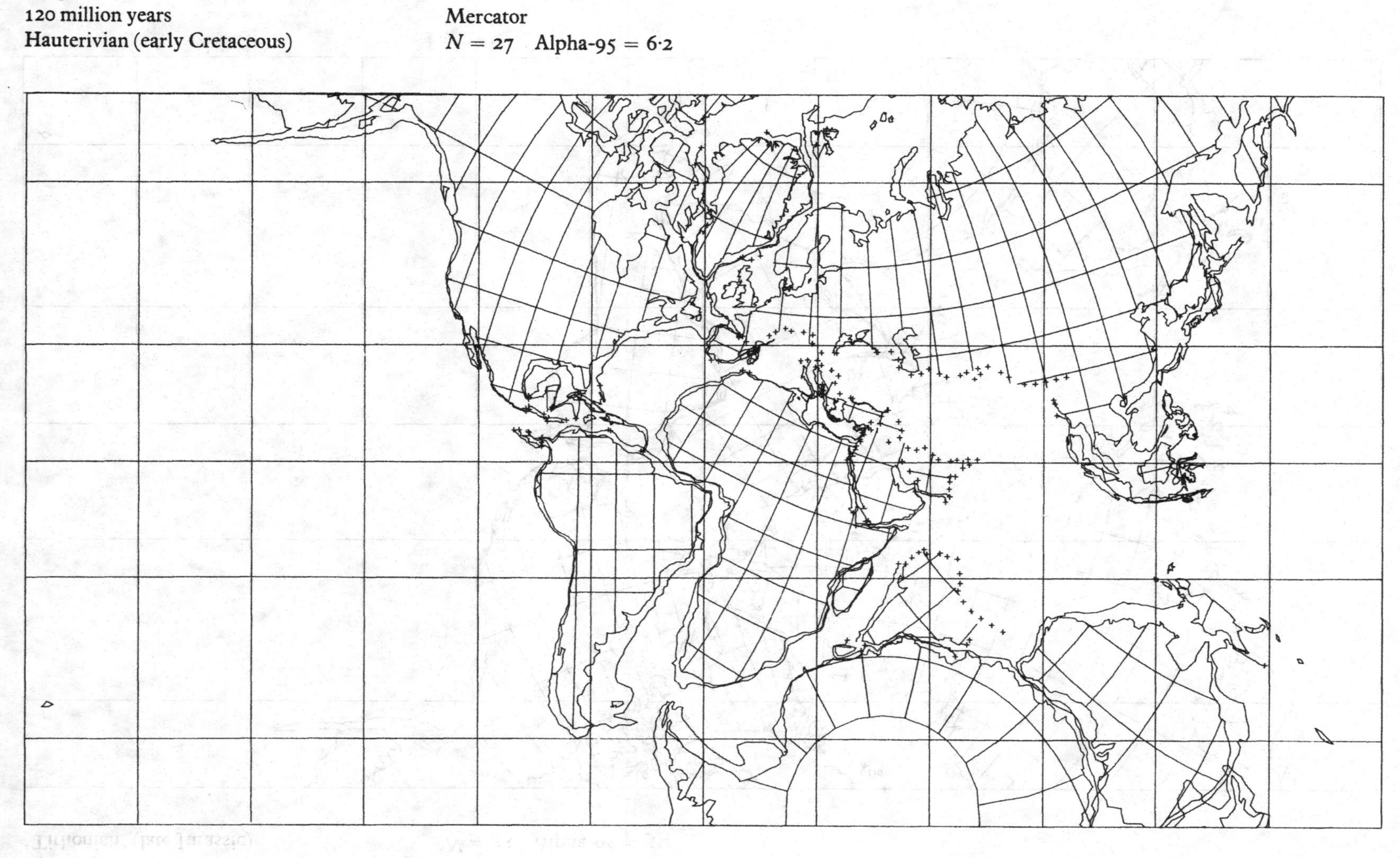

Map 9
140 million years
'Tithonian' (late Jurassic)

Mercator
$N = 33$ Alpha-95 = 5·4

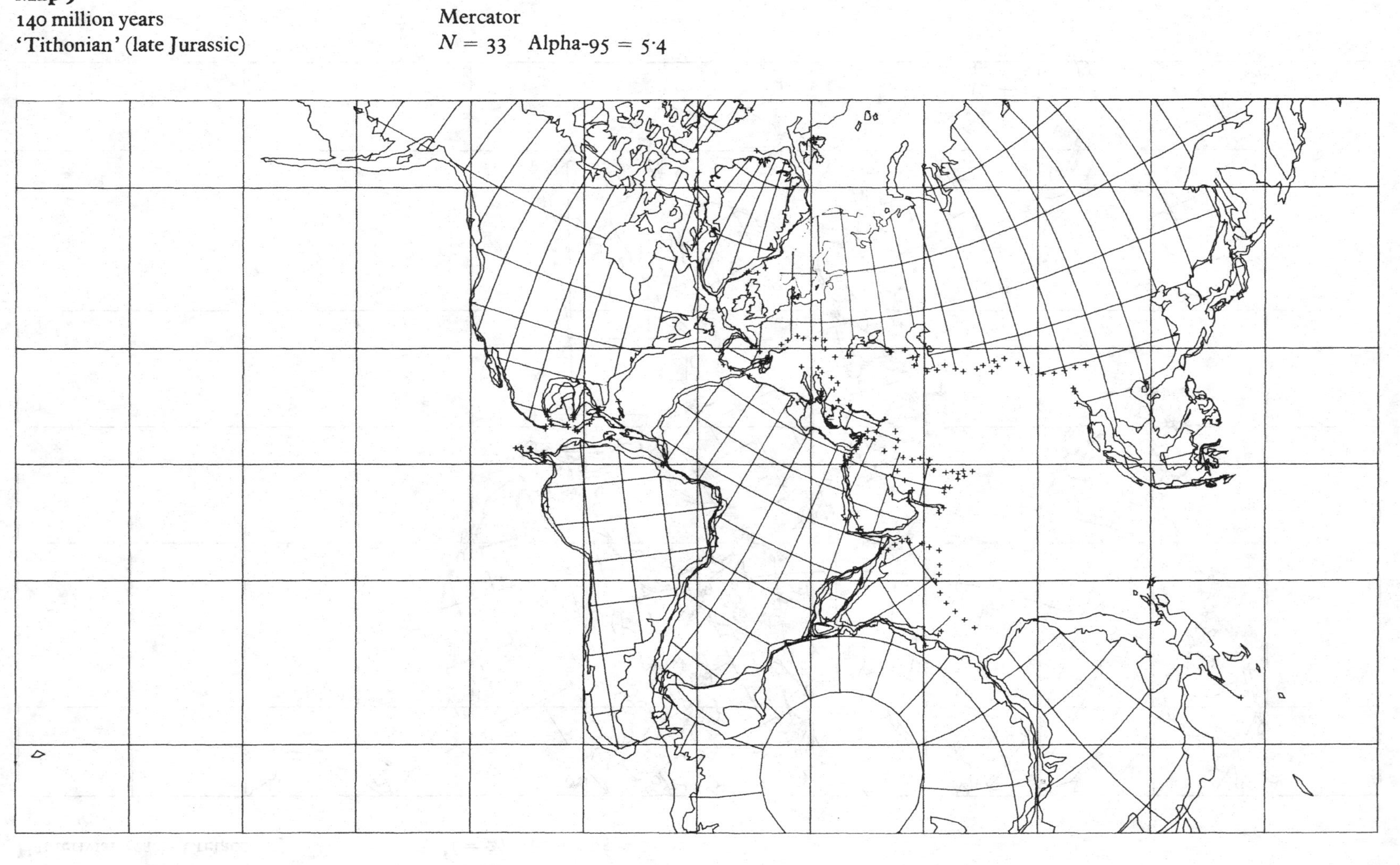

Map 10
160 million years
Callovian (mid-Jurassic)

Mercator
$N = 18$ Alpha-95 = 8·7

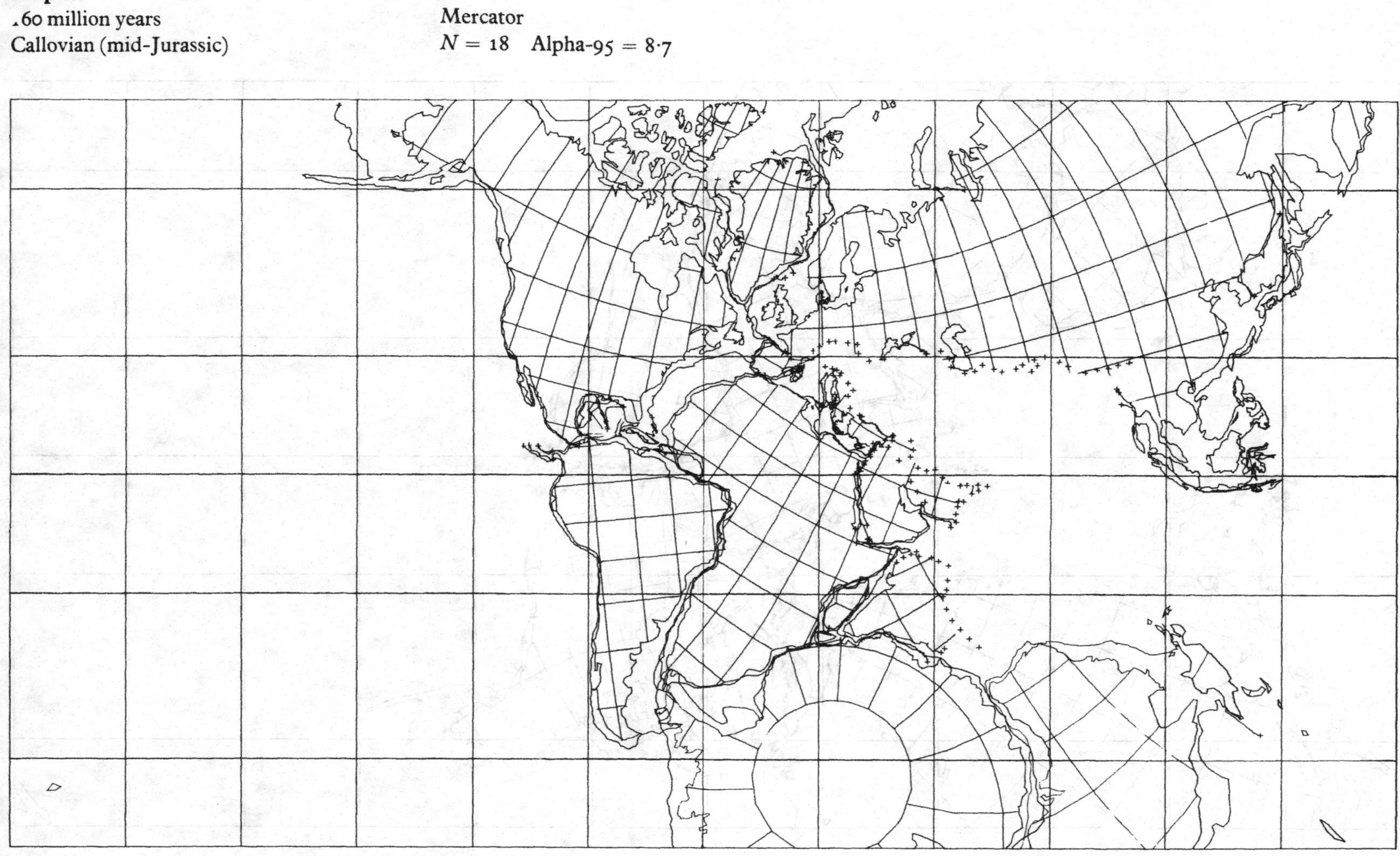

Map 11
180 million years
Pliensbachian (early Jurassic)

Mercator
$N = 32$ Alpha-95 = 7·5

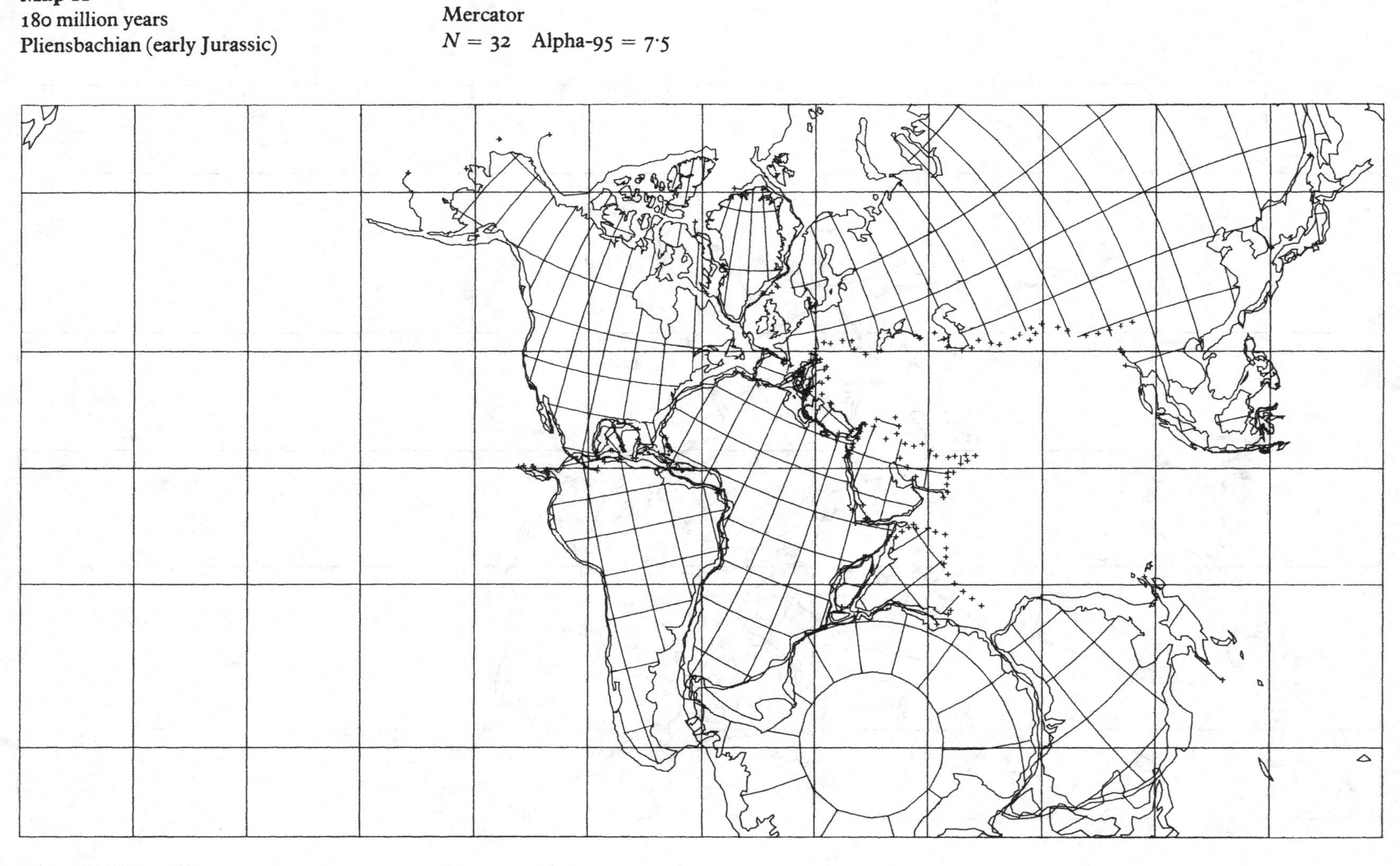

Map 12
200 million years
approx. Rhaetian (latest Triassic)

Mercator
$N = 36$ Alpha-95 = 6·1

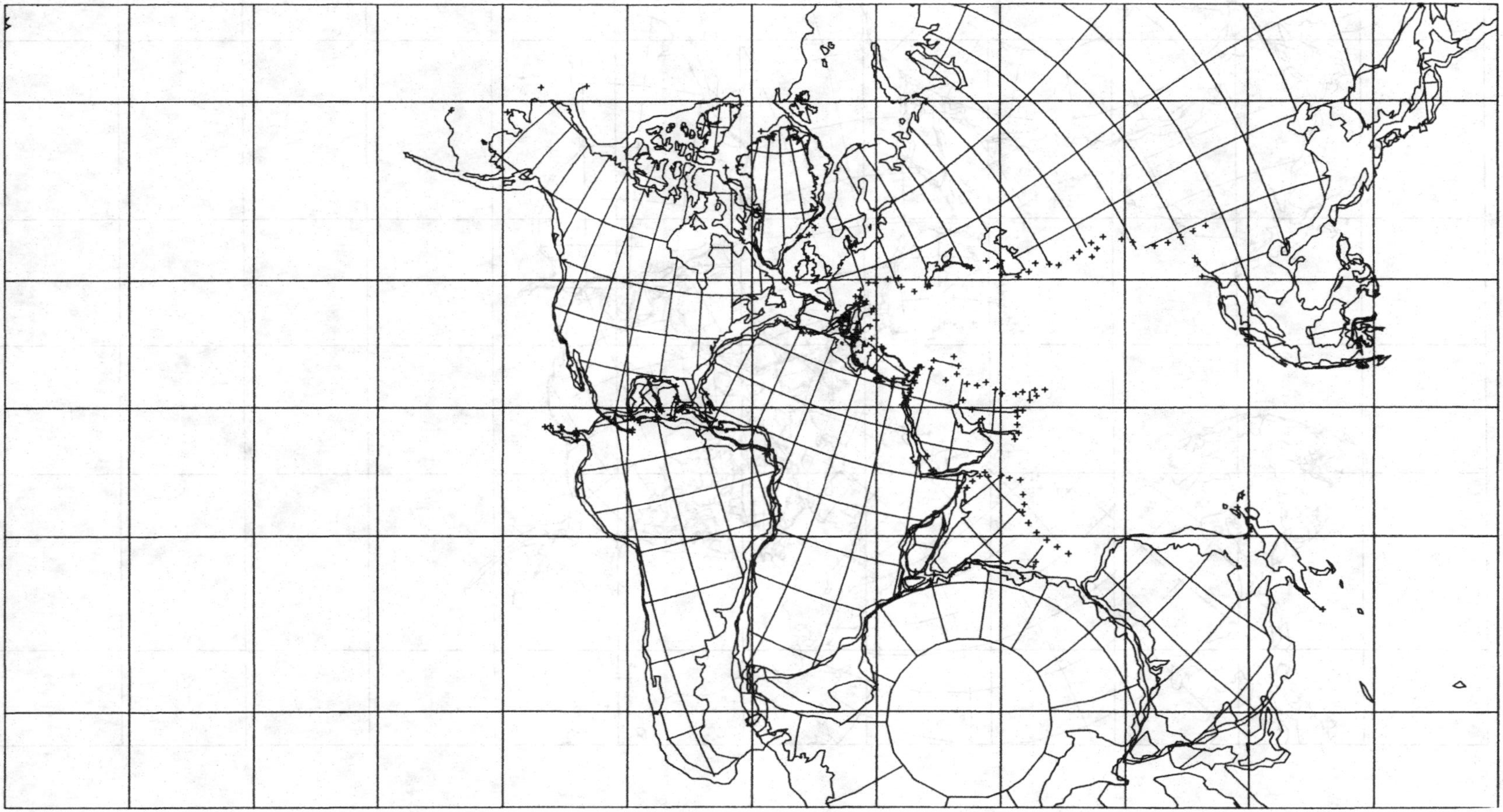

Map 13
220 million years
early Triassic

Mercator
$N = 67$ Alpha-95 = 4·5

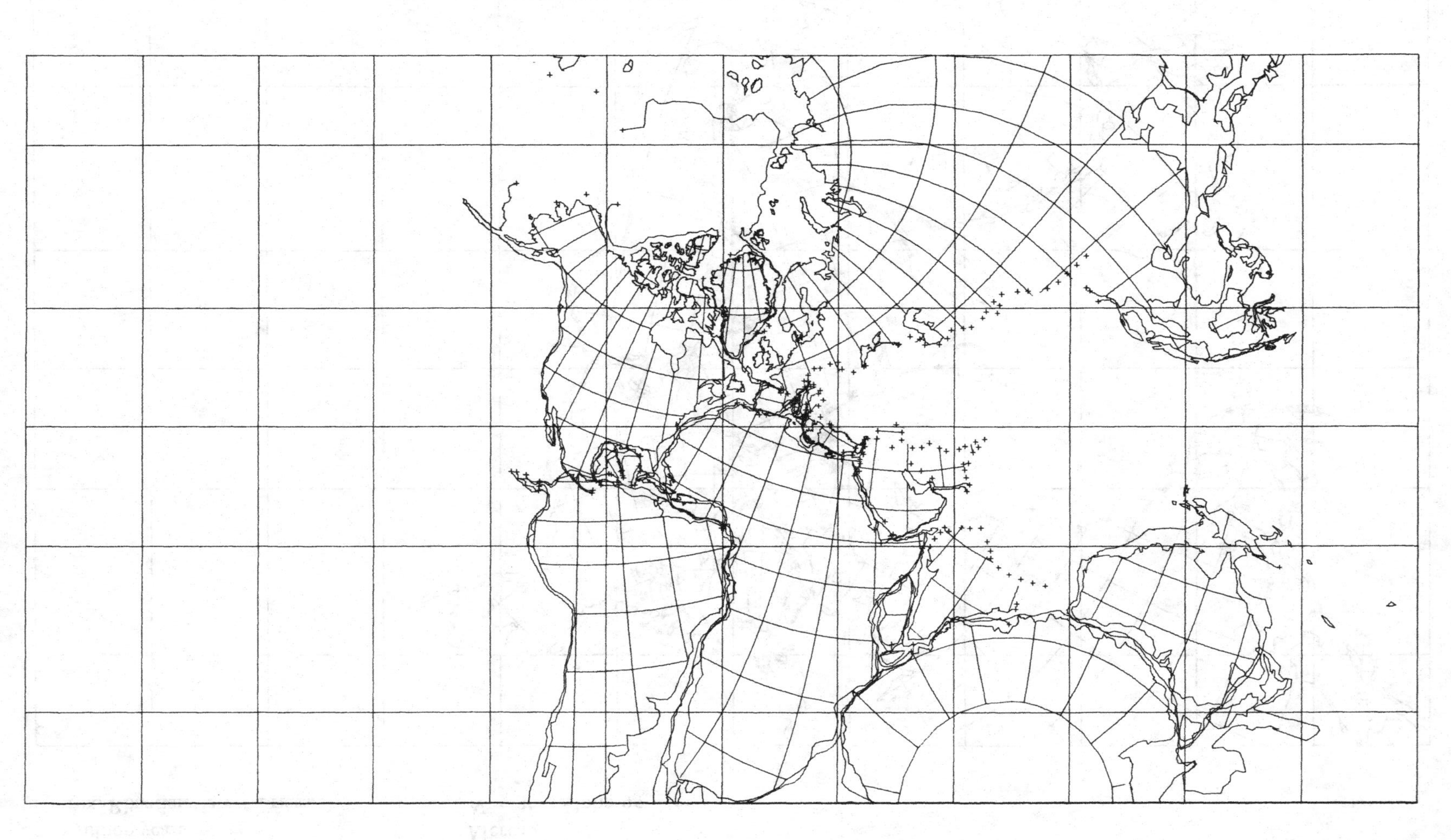

Map 14
Present day

North polar stereographic

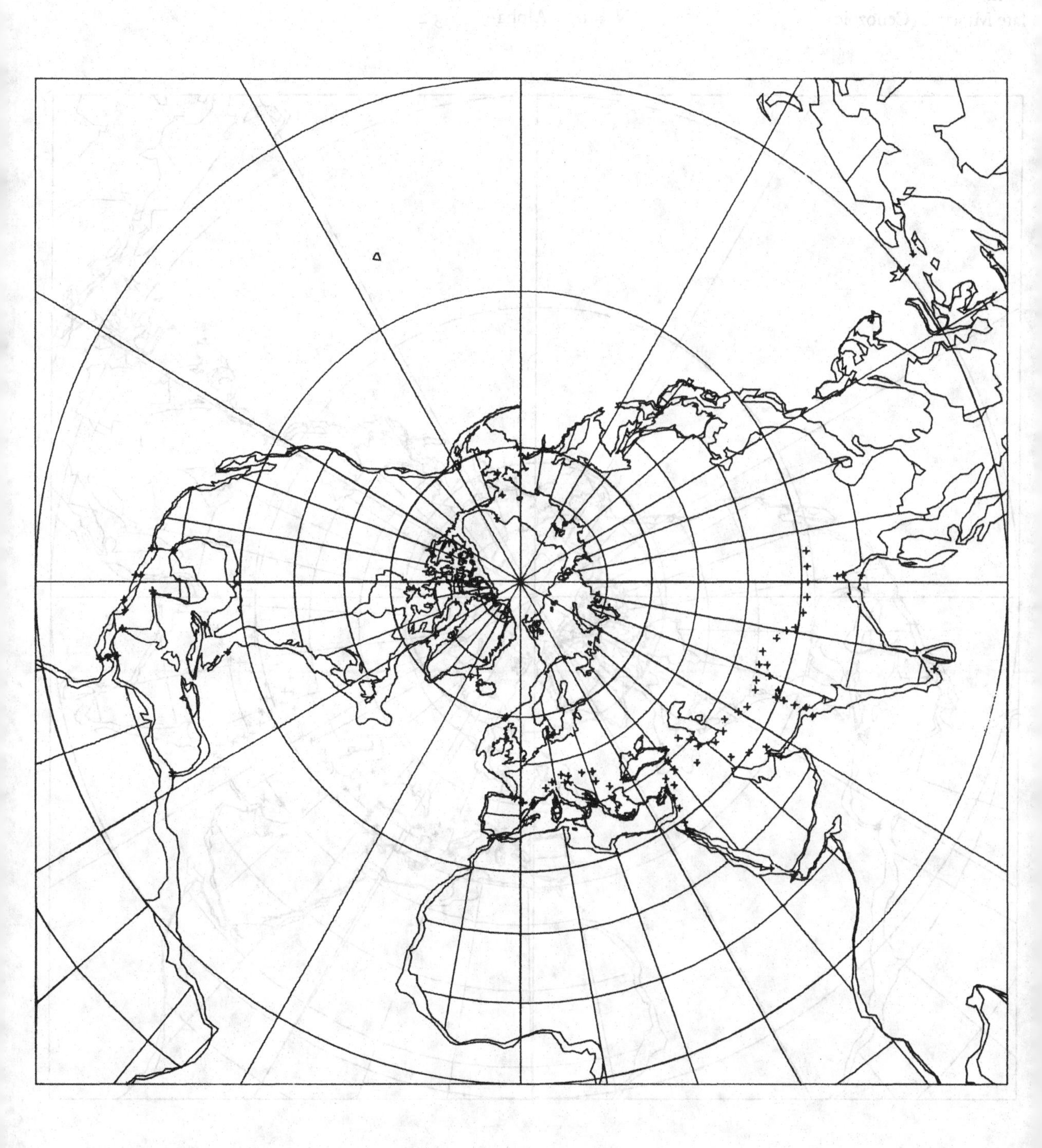

Map 15
10 million years
late Miocene (Cenozoic)

North polar stereographic
$N = 87$ Alpha-95 = 3·2

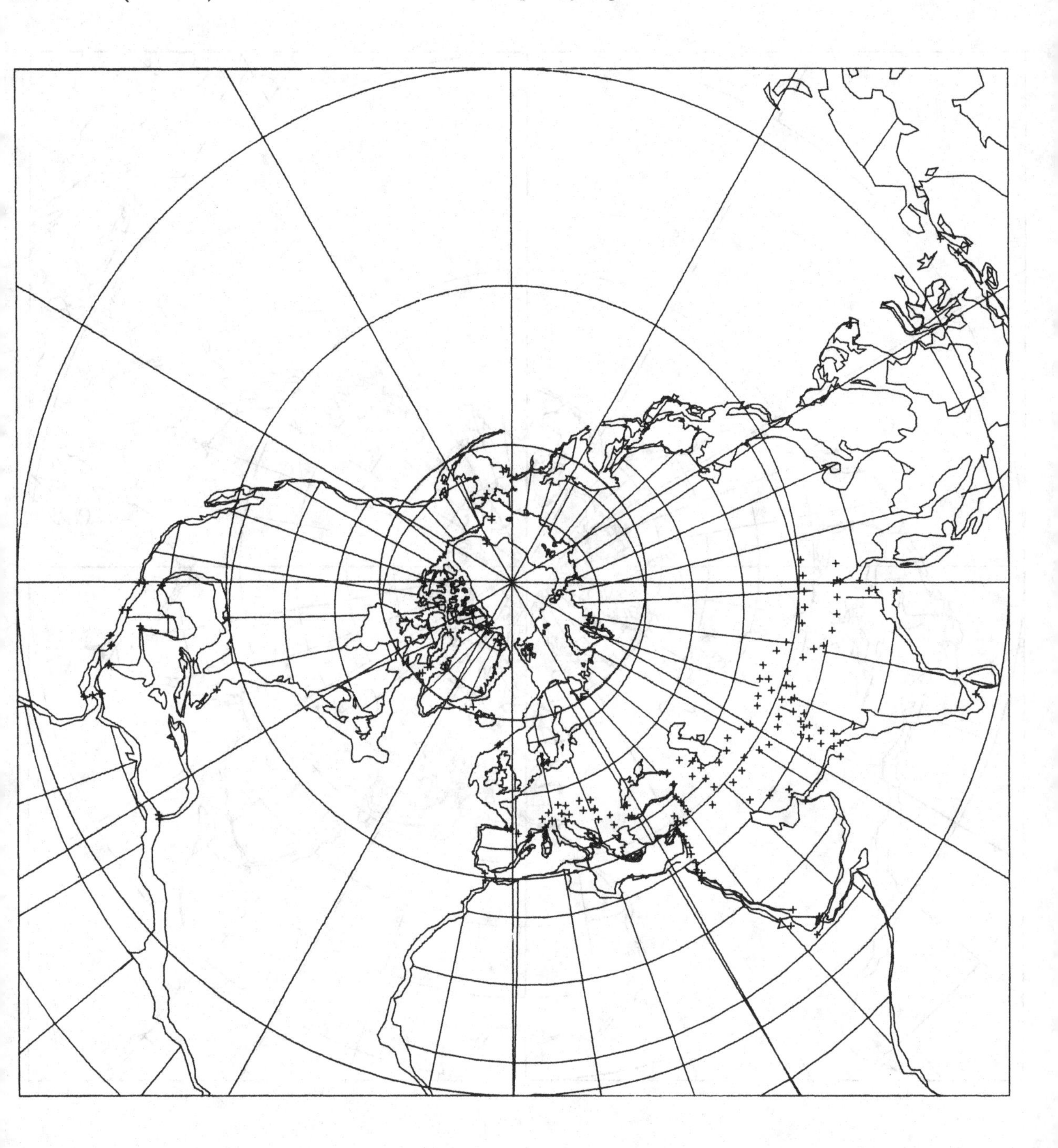

Map 16
20 million years
early Miocene (Cenozoic)

North polar stereographic
$N = 46$ Alpha-95 = 5·7

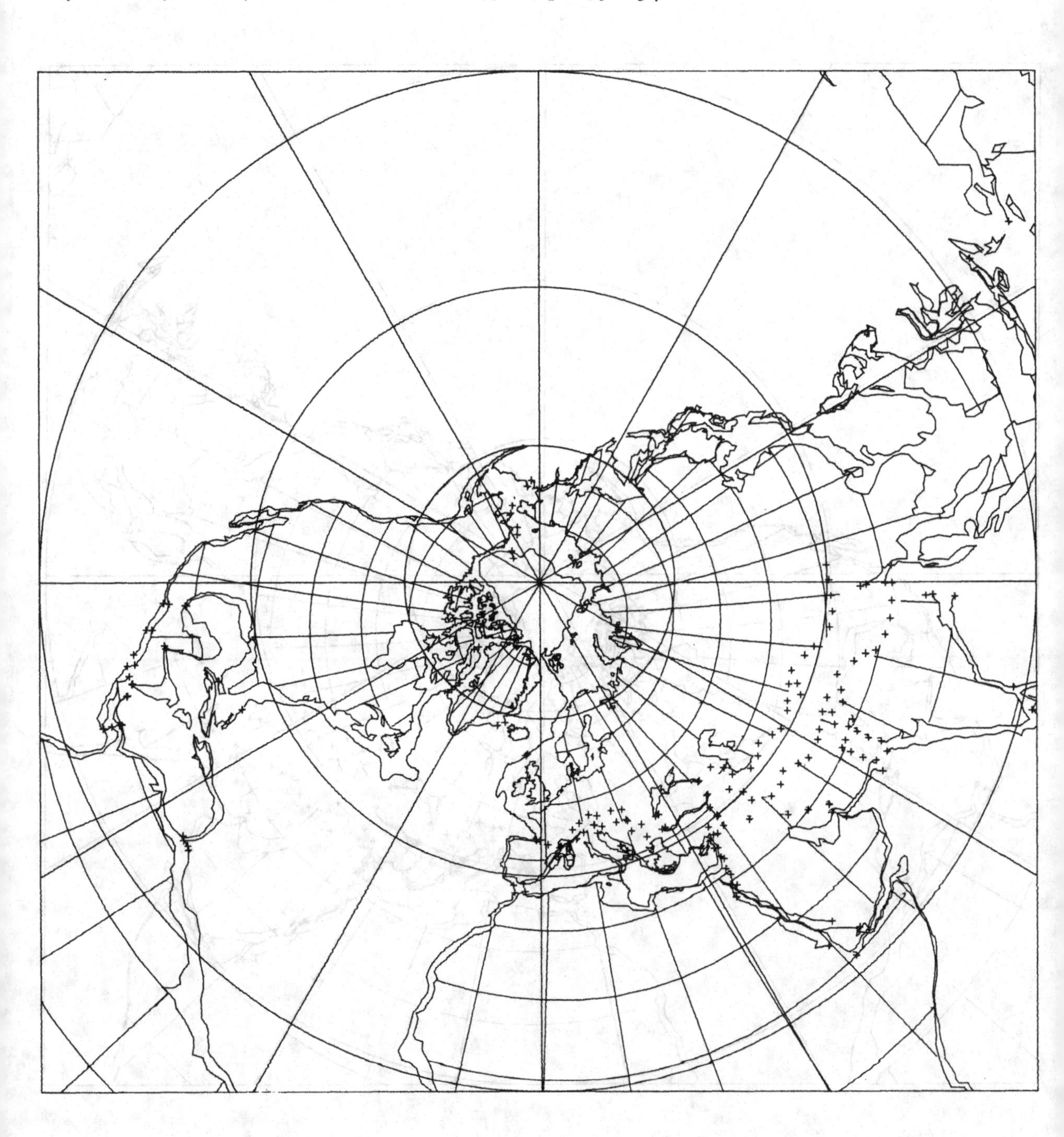

Map 17

40 million years
late Eocene (Cenozoic)

North polar stereographic
$N = 28$ Alpha-95 = 4·5

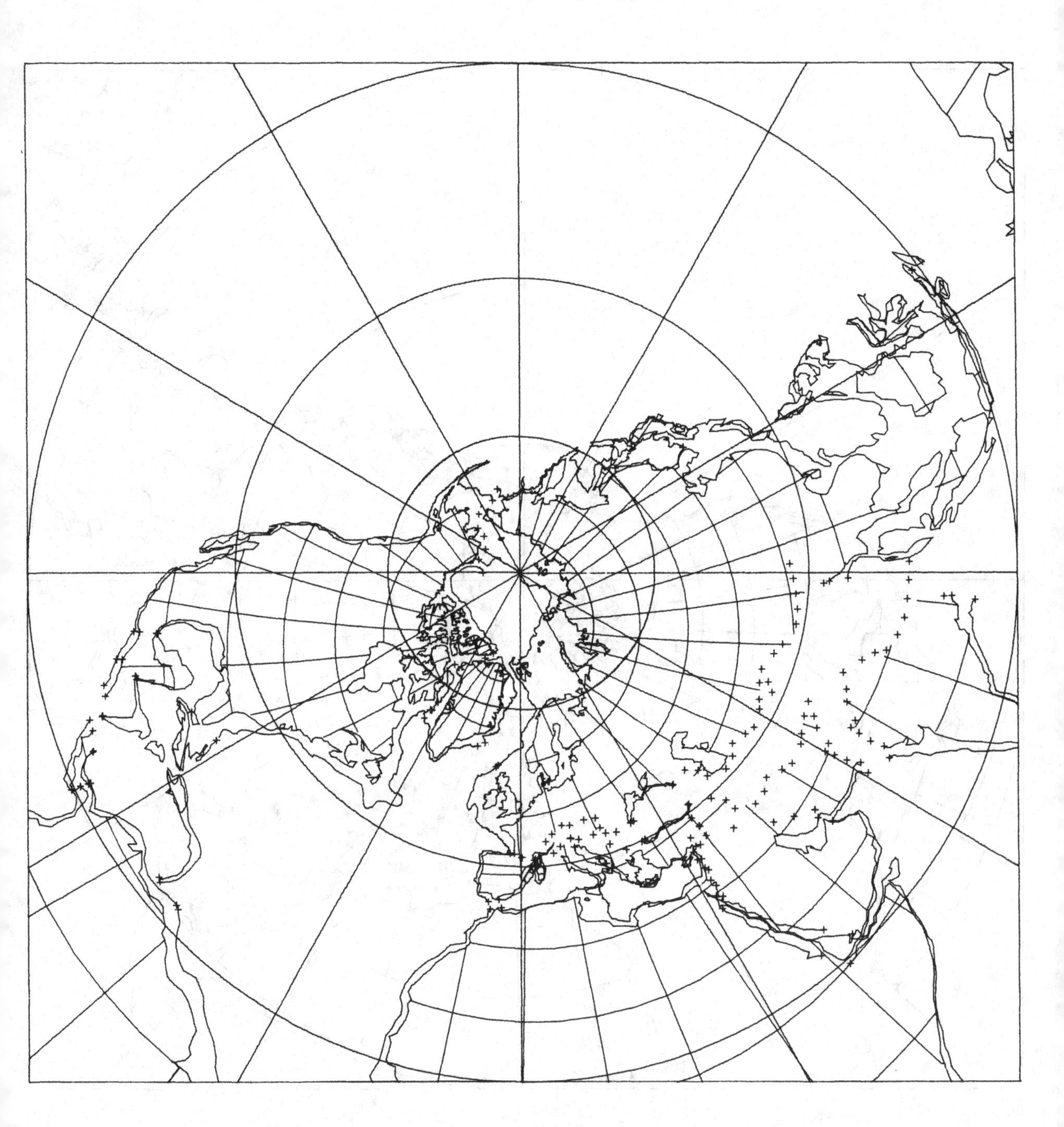

Map 18
60 million years
Paleocene (Cenozoic)

North polar stereographic
$N = 43$ Alpha-95 = 4·7

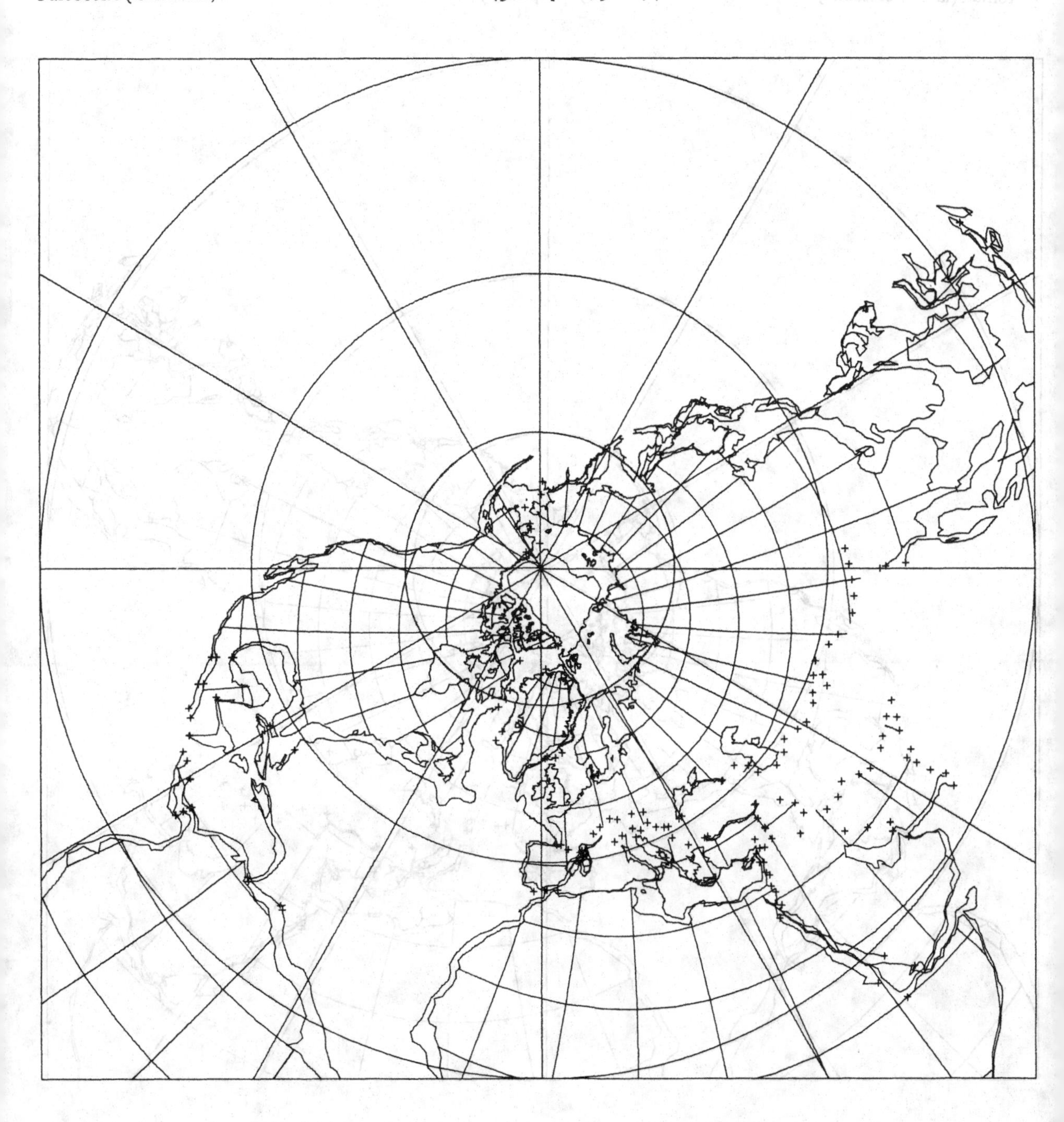

Map 19

80 million years

Santonian (late Cretaceous)

North polar stereographic

$N = 25$ Alpha-95 = 6·0

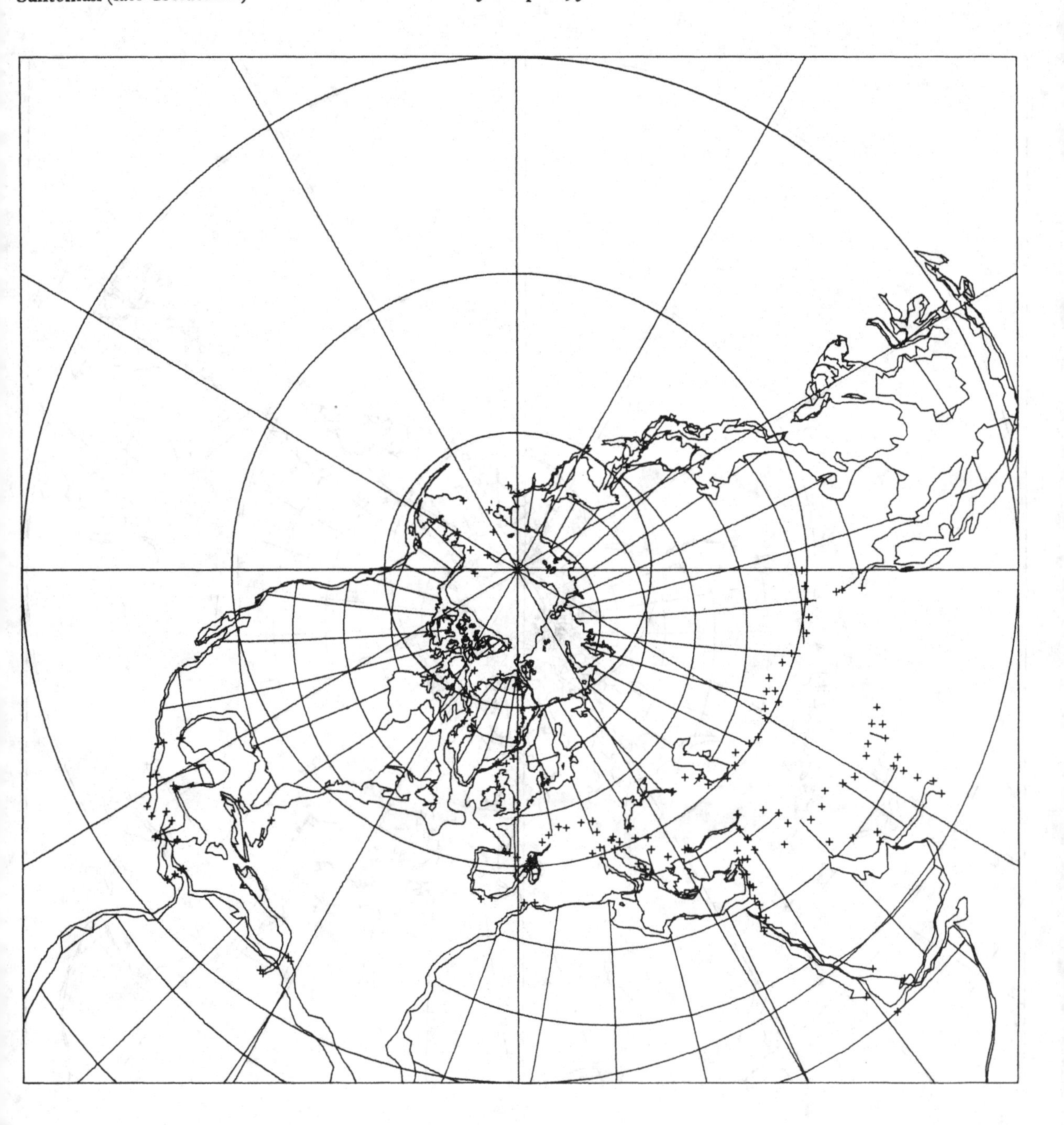

Map 20
100 million years
earliest Cenomanian (mid-Cretaceous)

North polar stereographic
$N = 43$ Alpha-95 = 5·2

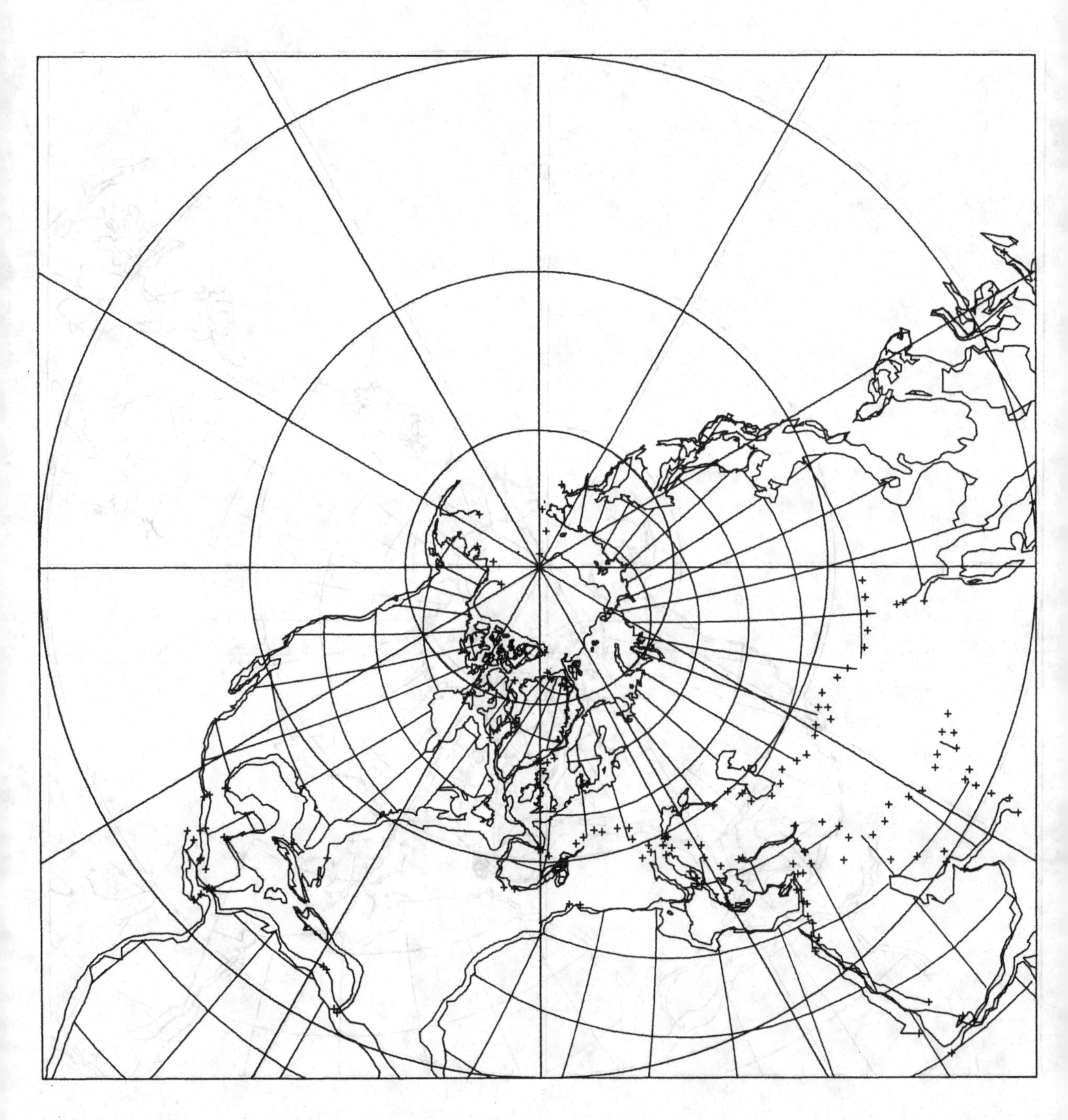

Map 21

120 million years

Hauterivian (early Cretaceous)

North polar stereographic

$N = 27$ Alpha-95 = 6·2

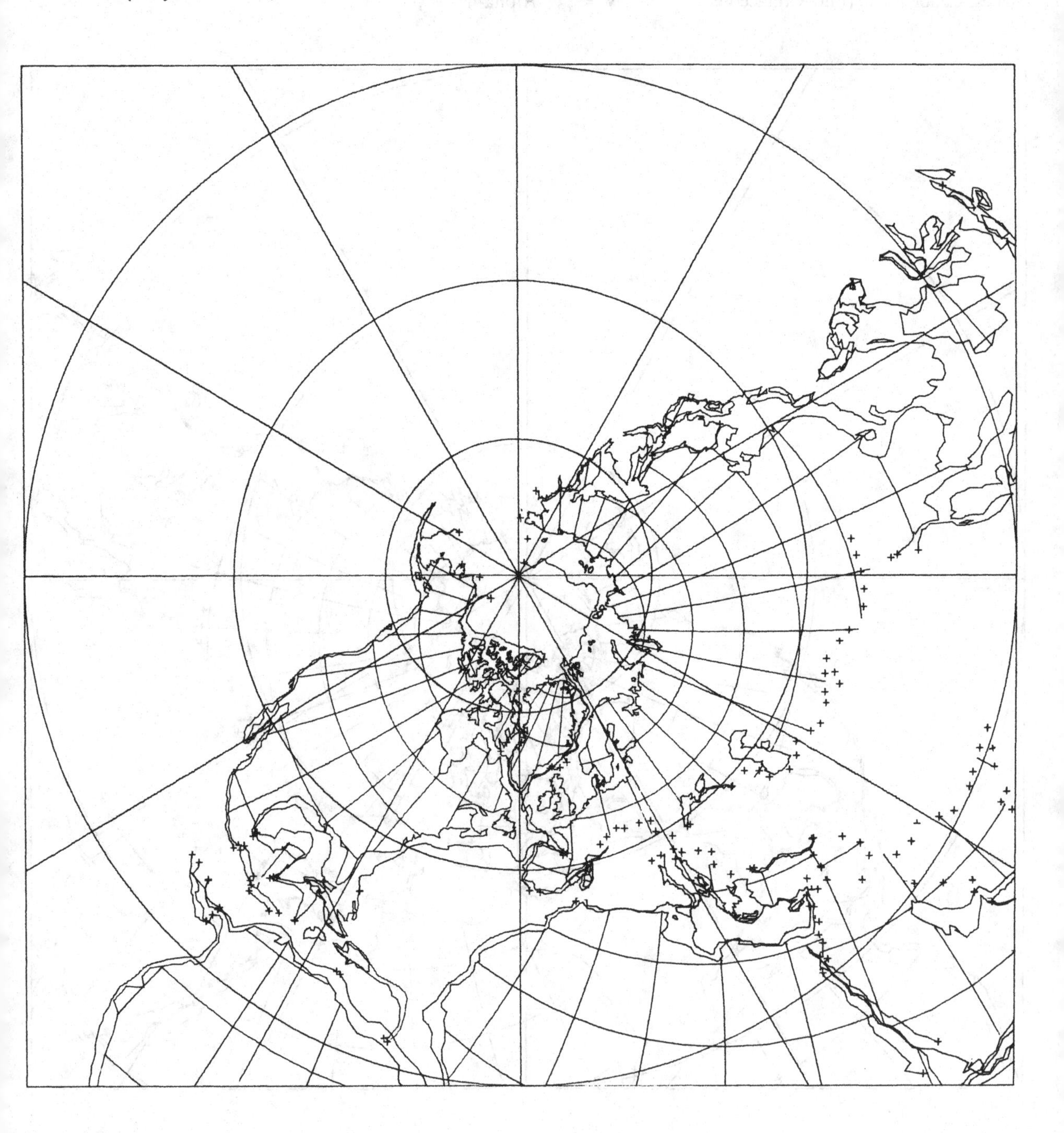

Map 22
140 million years
'Tithonian' (late Jurassic)

North polar stereographic
$N = 33$ Alpha-95 = 5·4

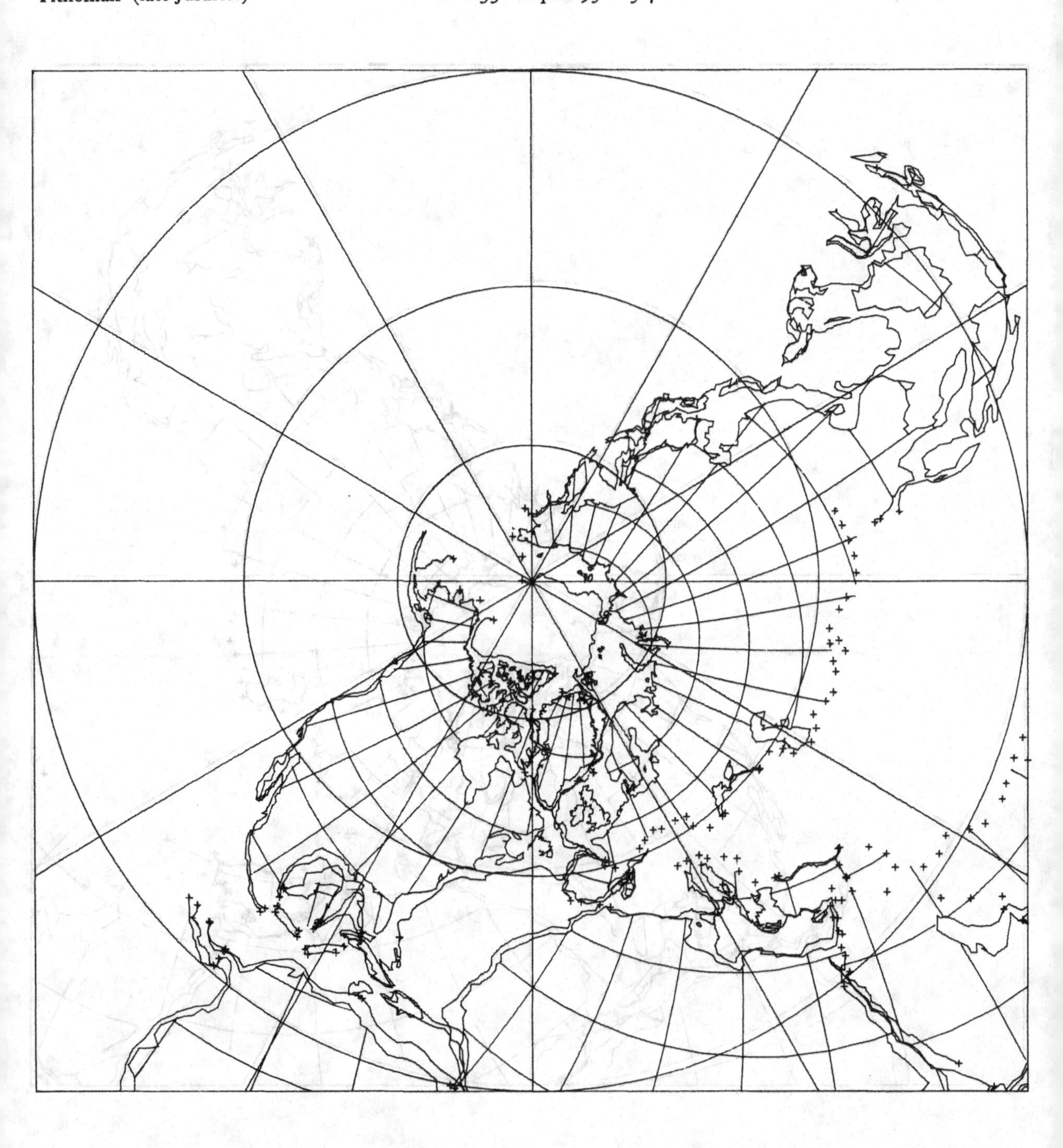

Map 23
160 million years
Callovian (mid-Jurassic)

North polar stereographic
$N = 18$ Alpha-95 $= 8{\cdot}7$

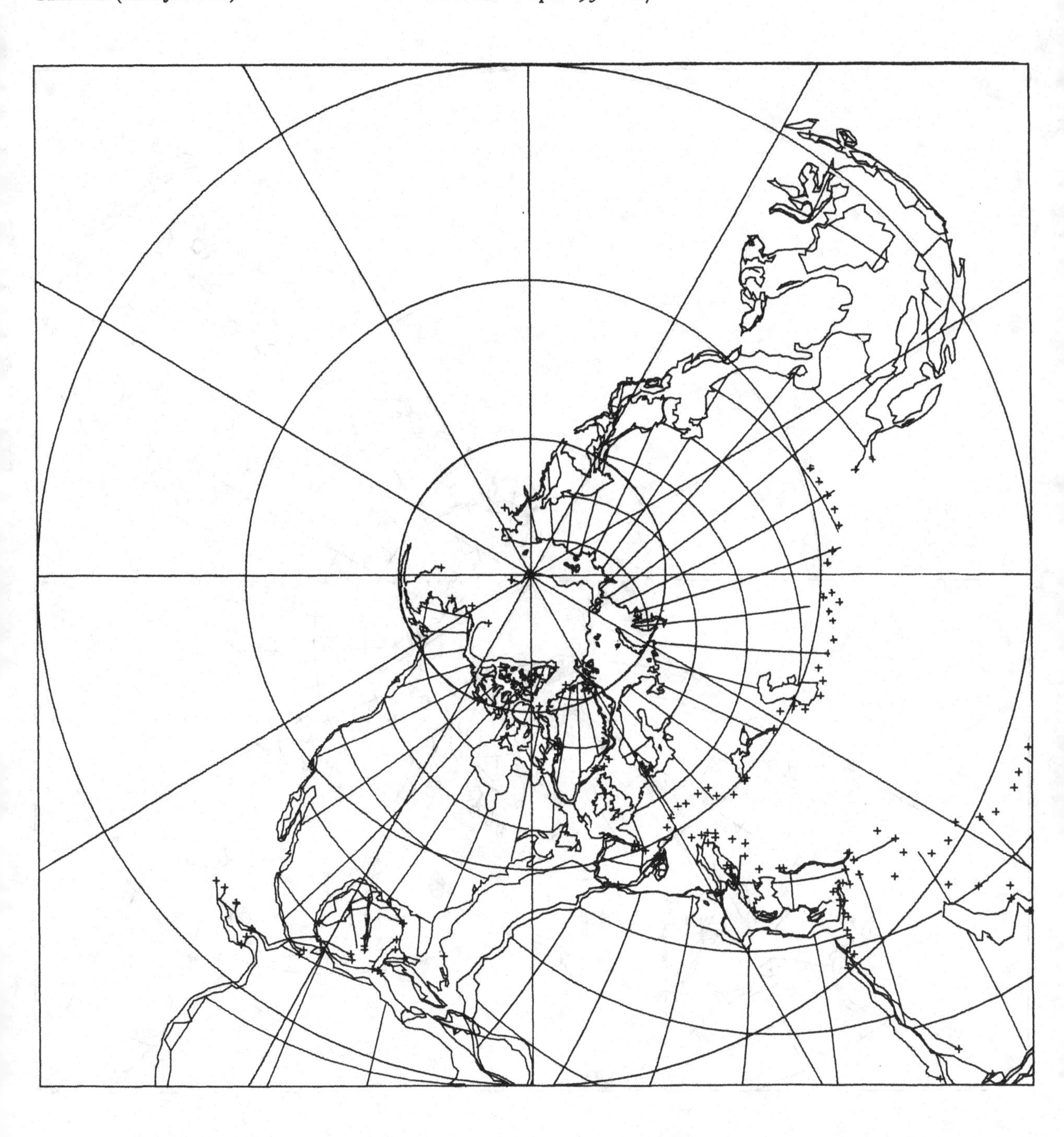

Map 24
180 million years
Pliensbachian (early Jurassic)

North polar stereographic
$N = 32$ Alpha-95 = 7·5

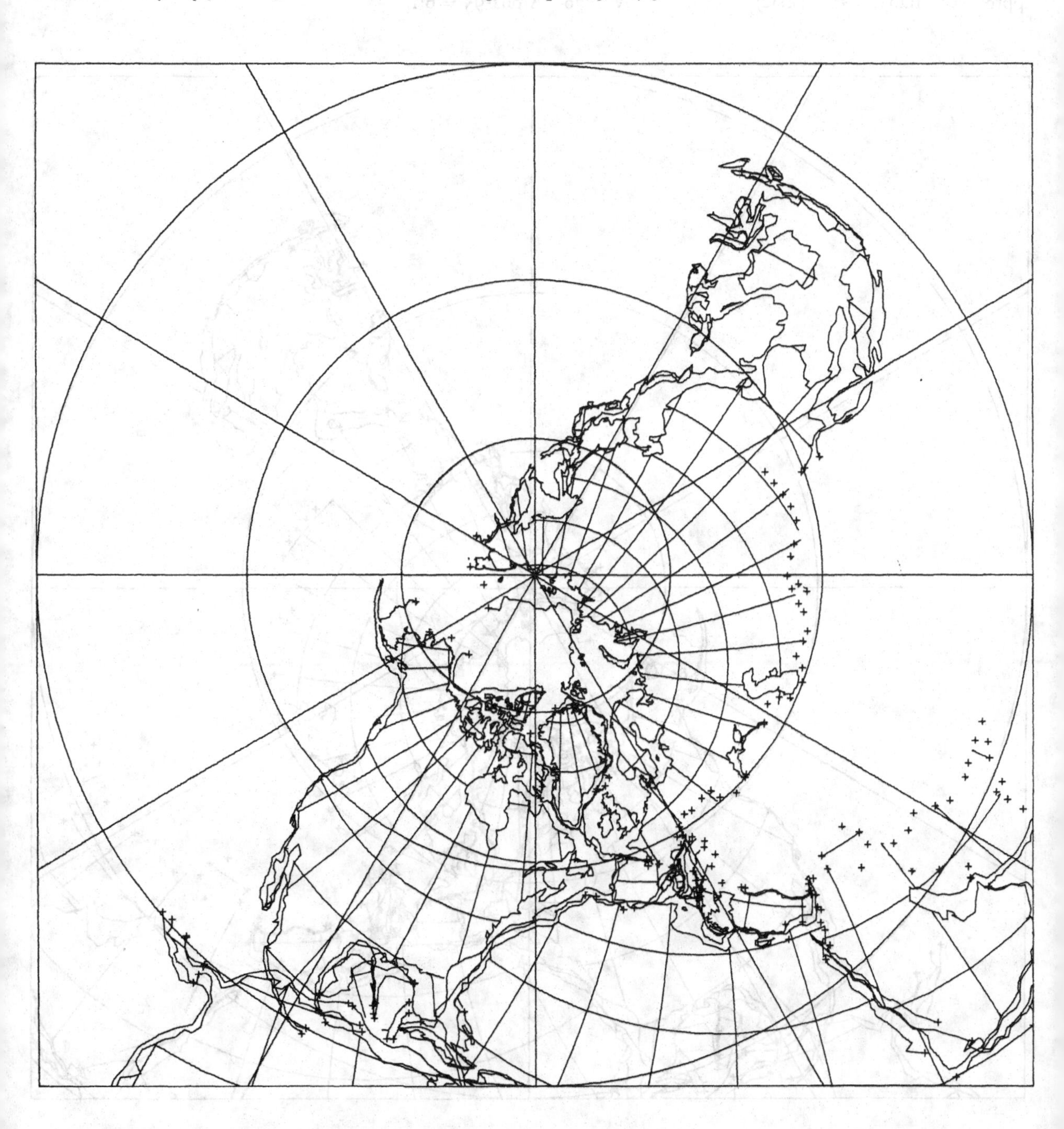

Map 25

200 million years
approx. Rhaetian (latest Triassic)

North polar stereographic
$N = 36$ Alpha-95 $= 6{\cdot}1$

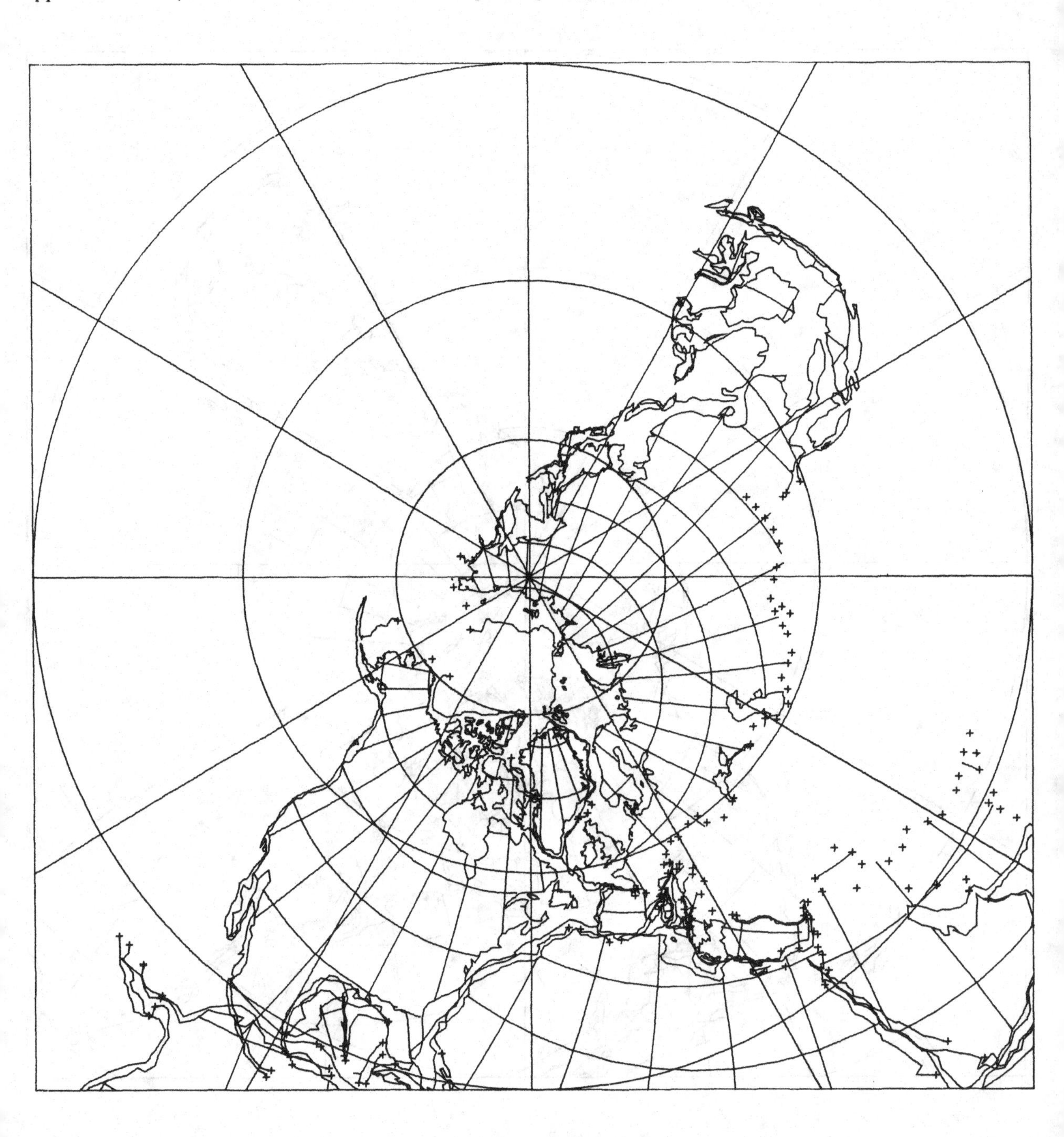

Map 26
220 million years
early Triassic

North polar stereographic
$N = 67$ Alpha-95 = 4·5

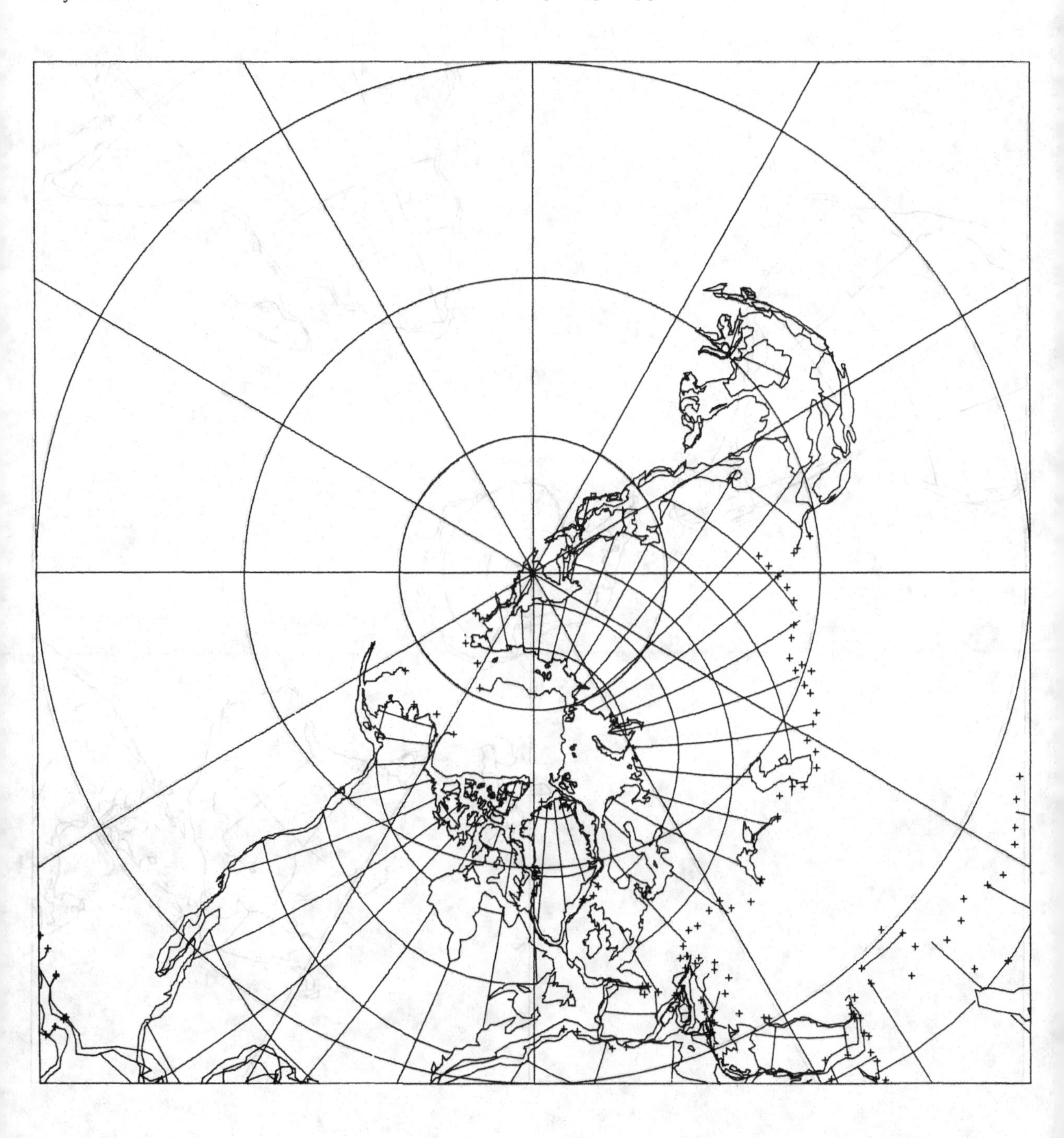

Map 27
Present day

South polar stereographic

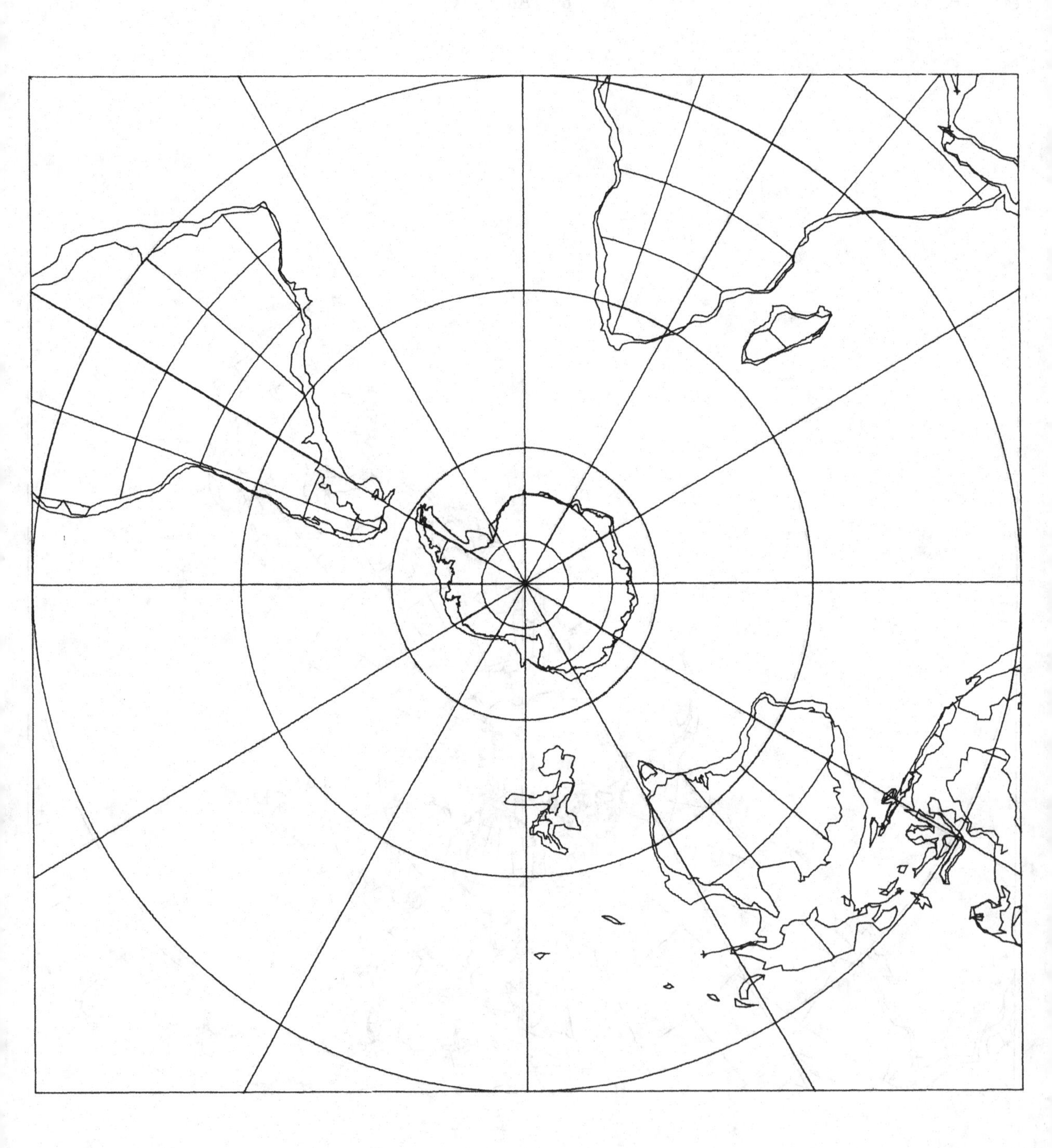

Map 28
10 million years
late Miocene (Cenozoic)

South polar stereographic
$N = 87$ Alpha-95 = 3·2

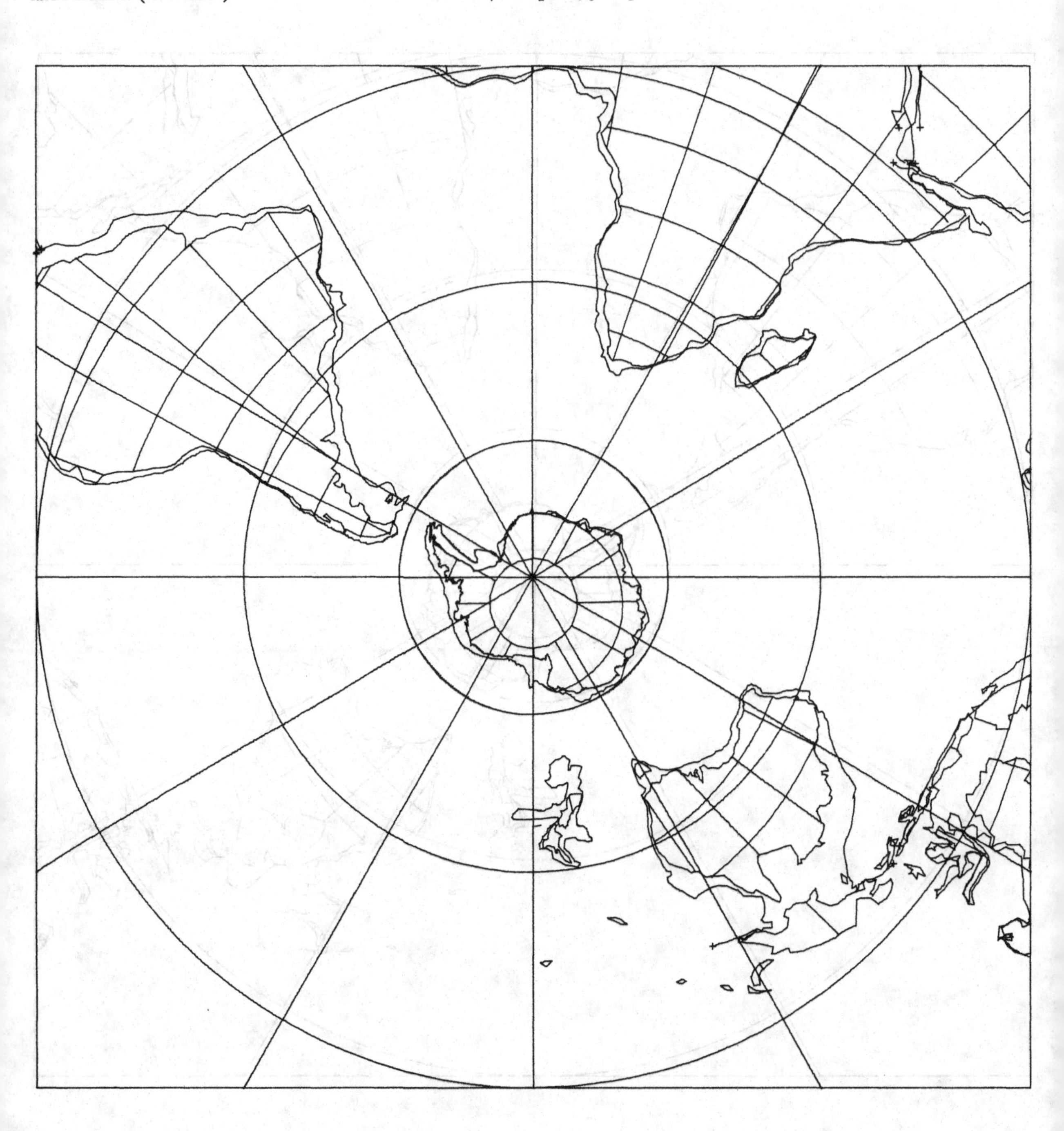

Map 29
20 million years
early Miocene (Cenozoic)

South polar stereographic
$N = 46$ Alpha-95 = 5·7

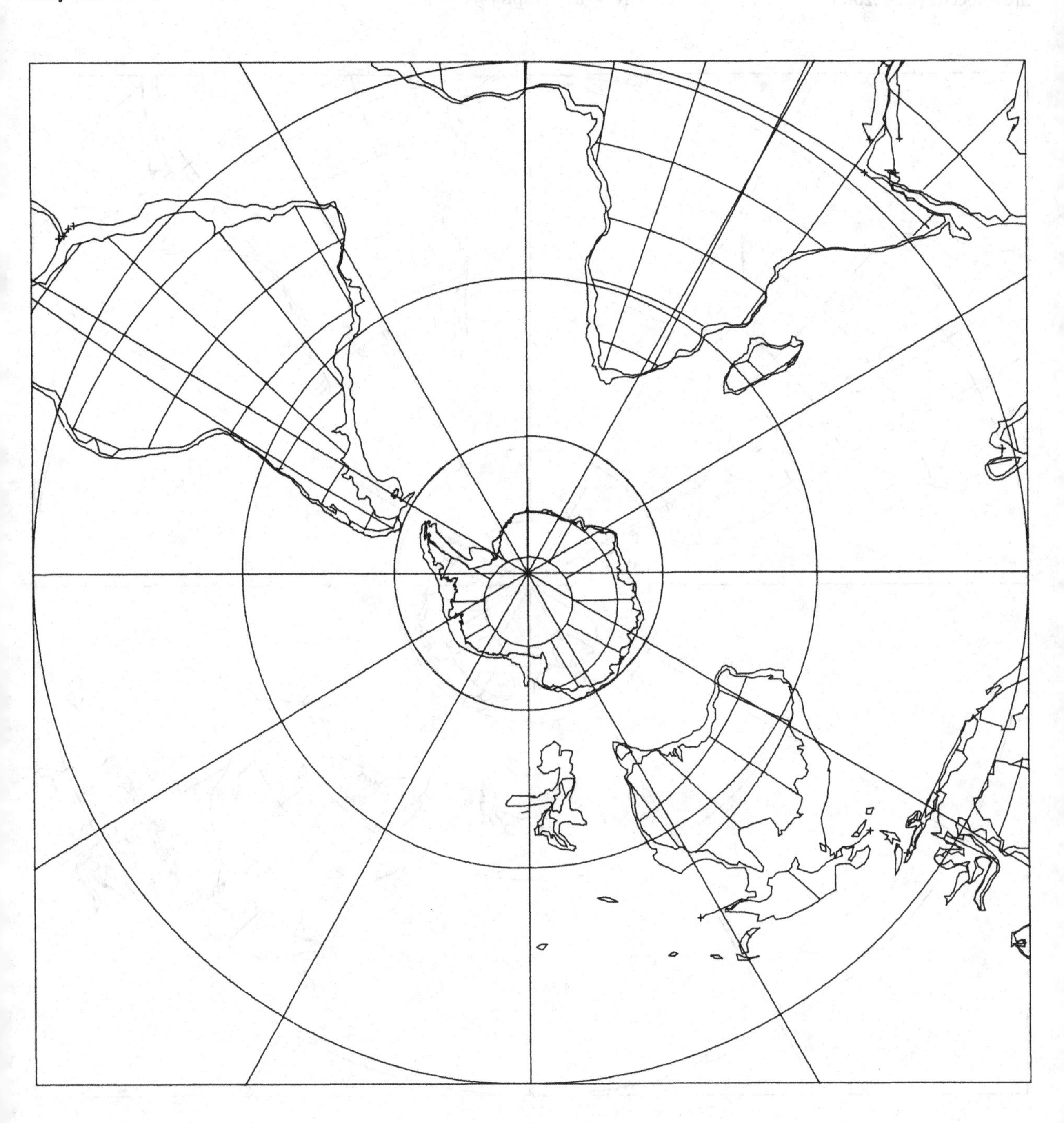

Map 30
40 million years
late Eocene (Cenozoic)

South polar stereographic
$N = 28$ Alpha-95 = 4·5

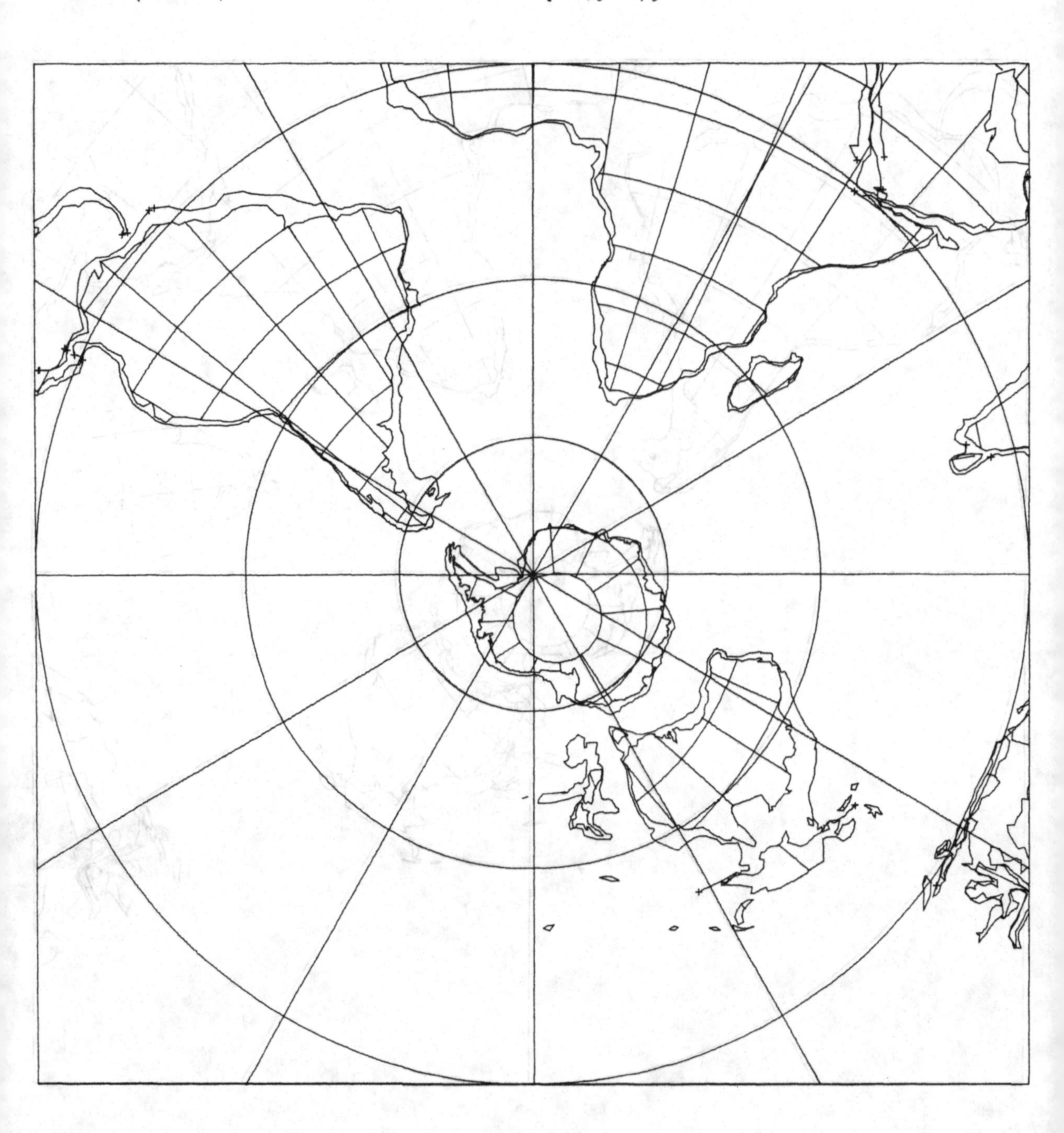

Map 31

60 million years

Paleocene (Cenozoic)

South polar stereographic

$N = 43$ Alpha-95 = 4·7

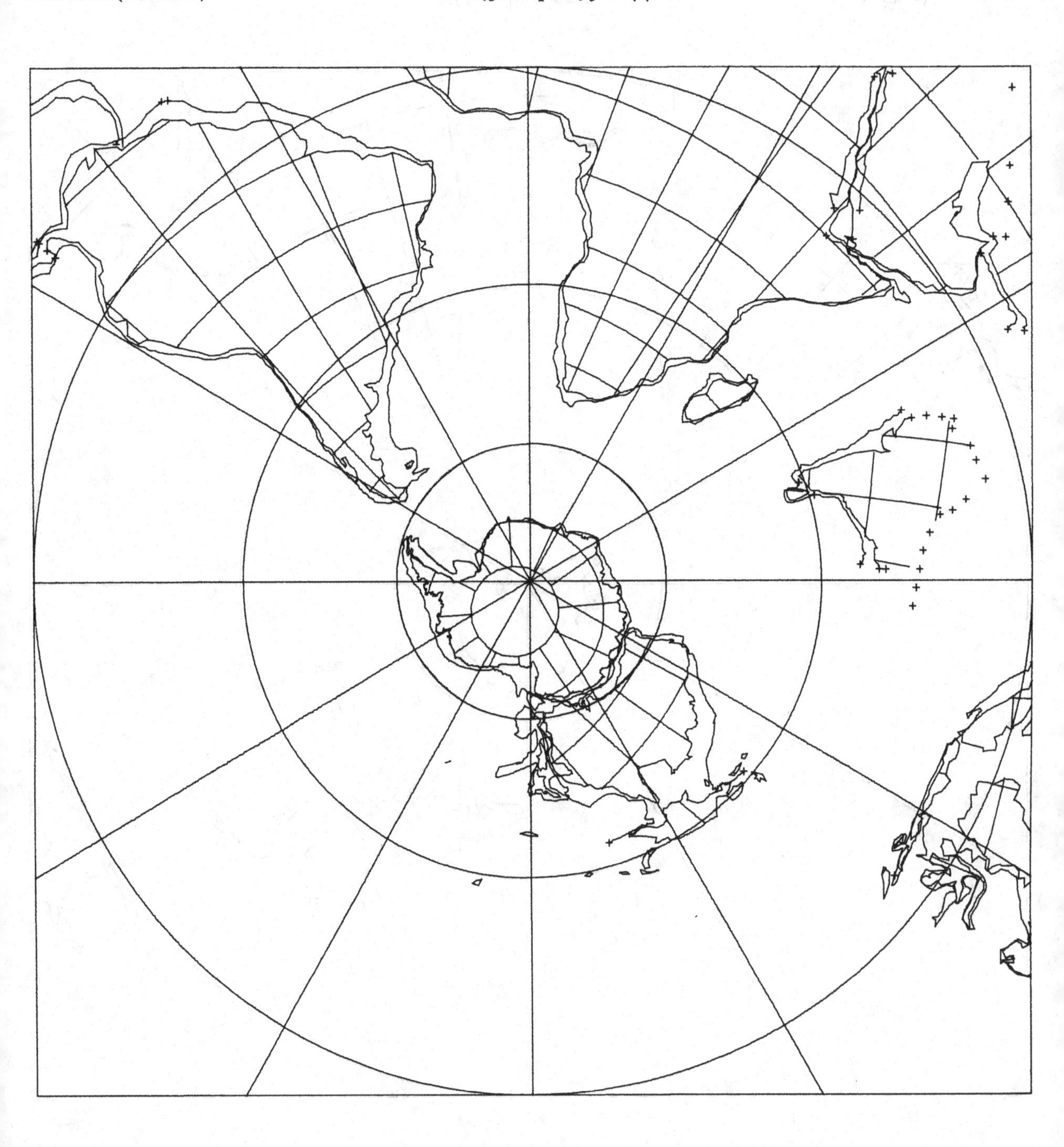

Map 32
80 million years
Santonian (late Cretaceous)

South polar stereographic
$N = 25$ Alpha-95 = 6·0

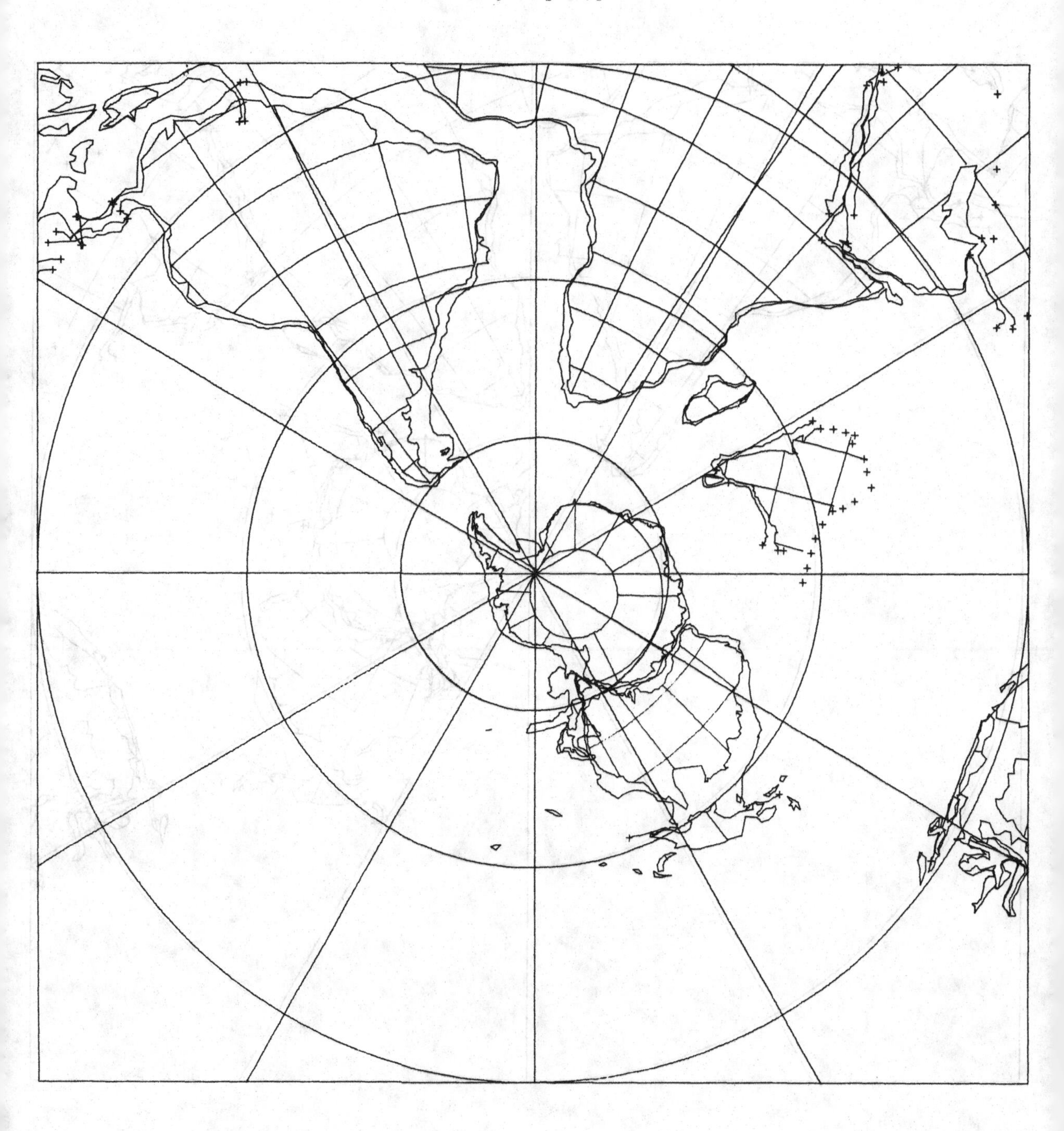

Map 33
100 million years
earliest Cenomanian (mid-Cretaceous)

South polar stereographic
$N = 43$ Alpha-95 = 5·2

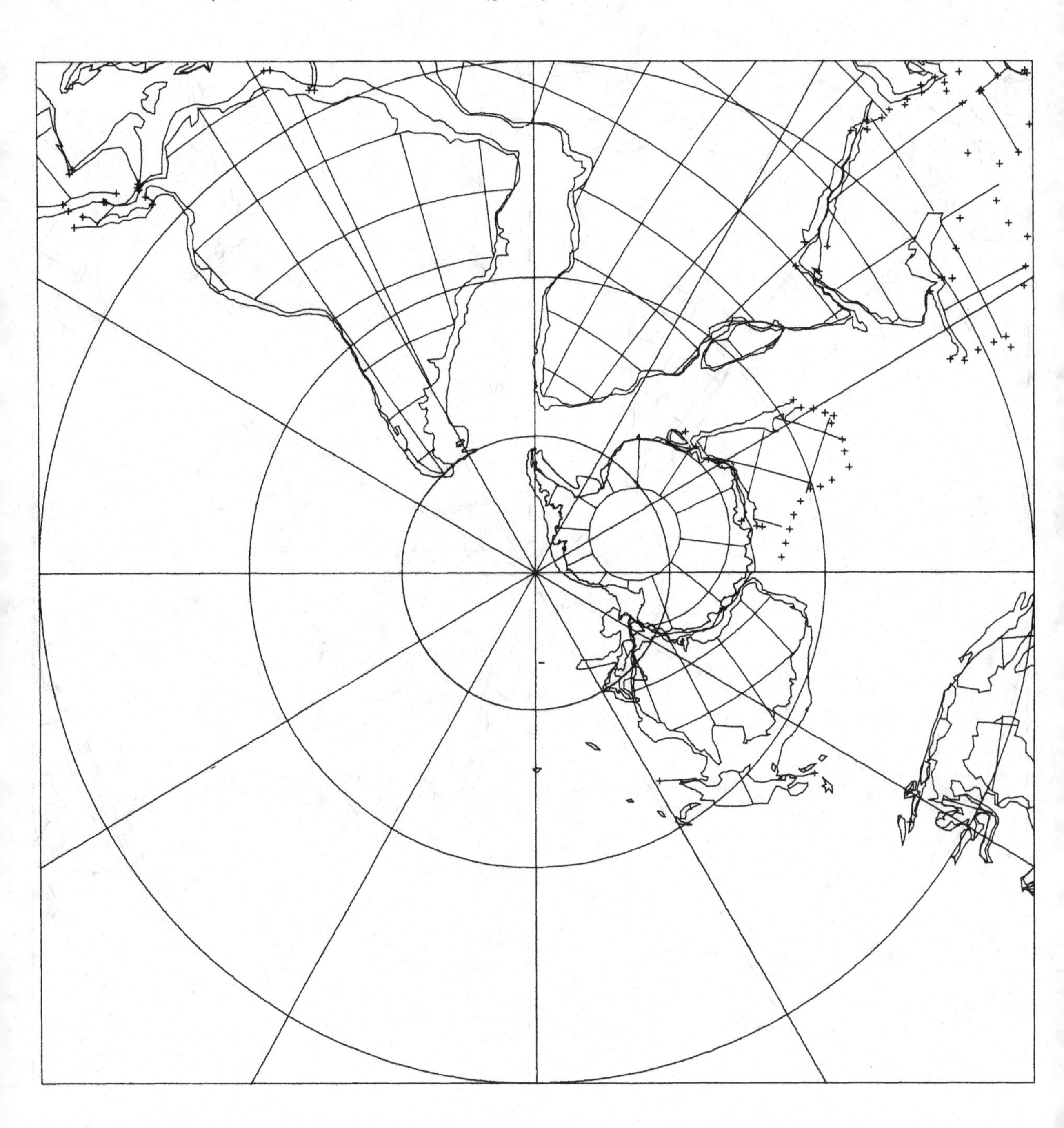

Map 34

120 million years

Hauterivian (early Cretaceous)

South polar stereographic

$N = 27$ Alpha-95 = 6·2

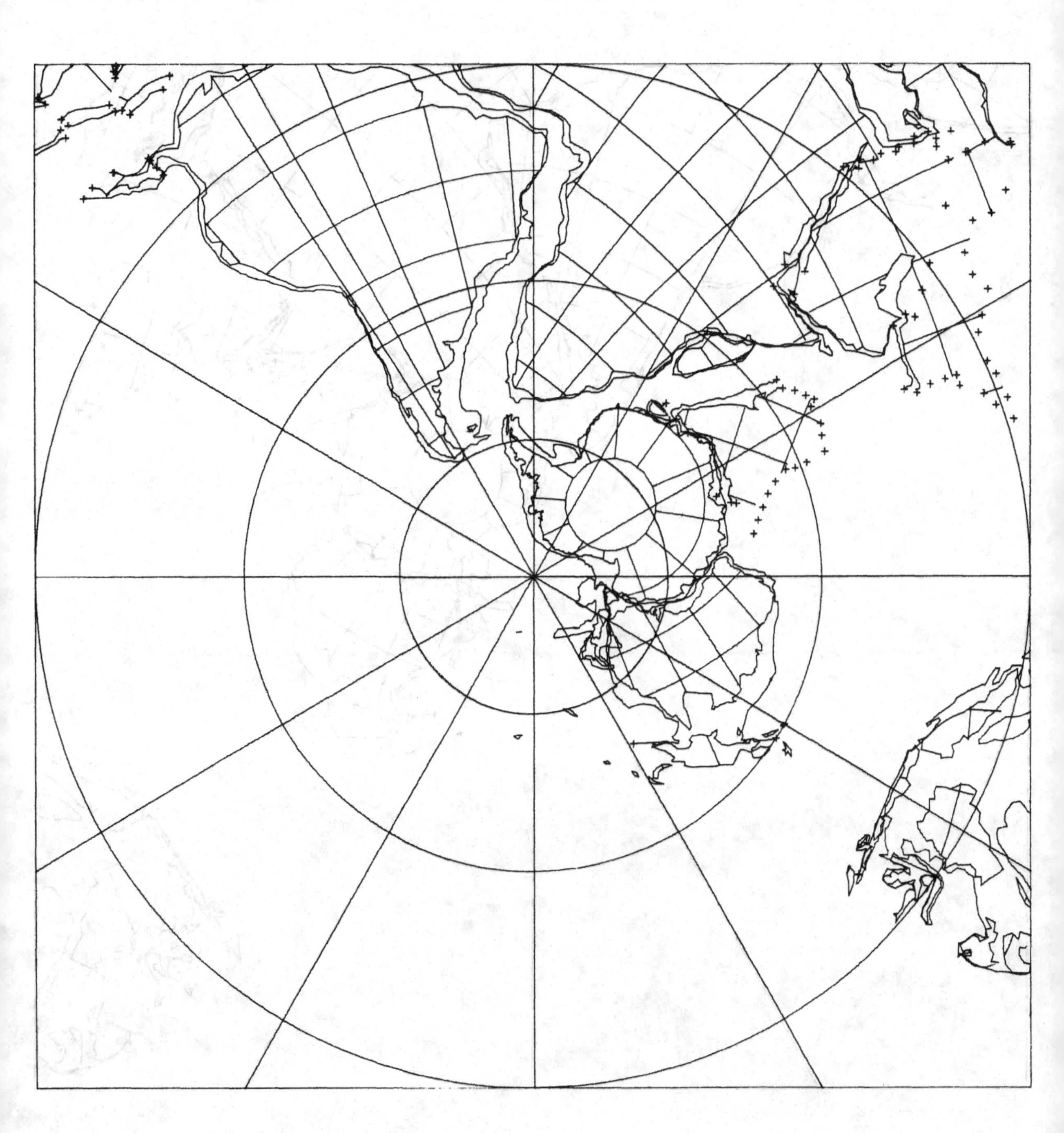

Map 35
140 million years
'Tithonian' (late Jurassic)

South polar stereographic
$N = 33$ Alpha-95 = 5.4

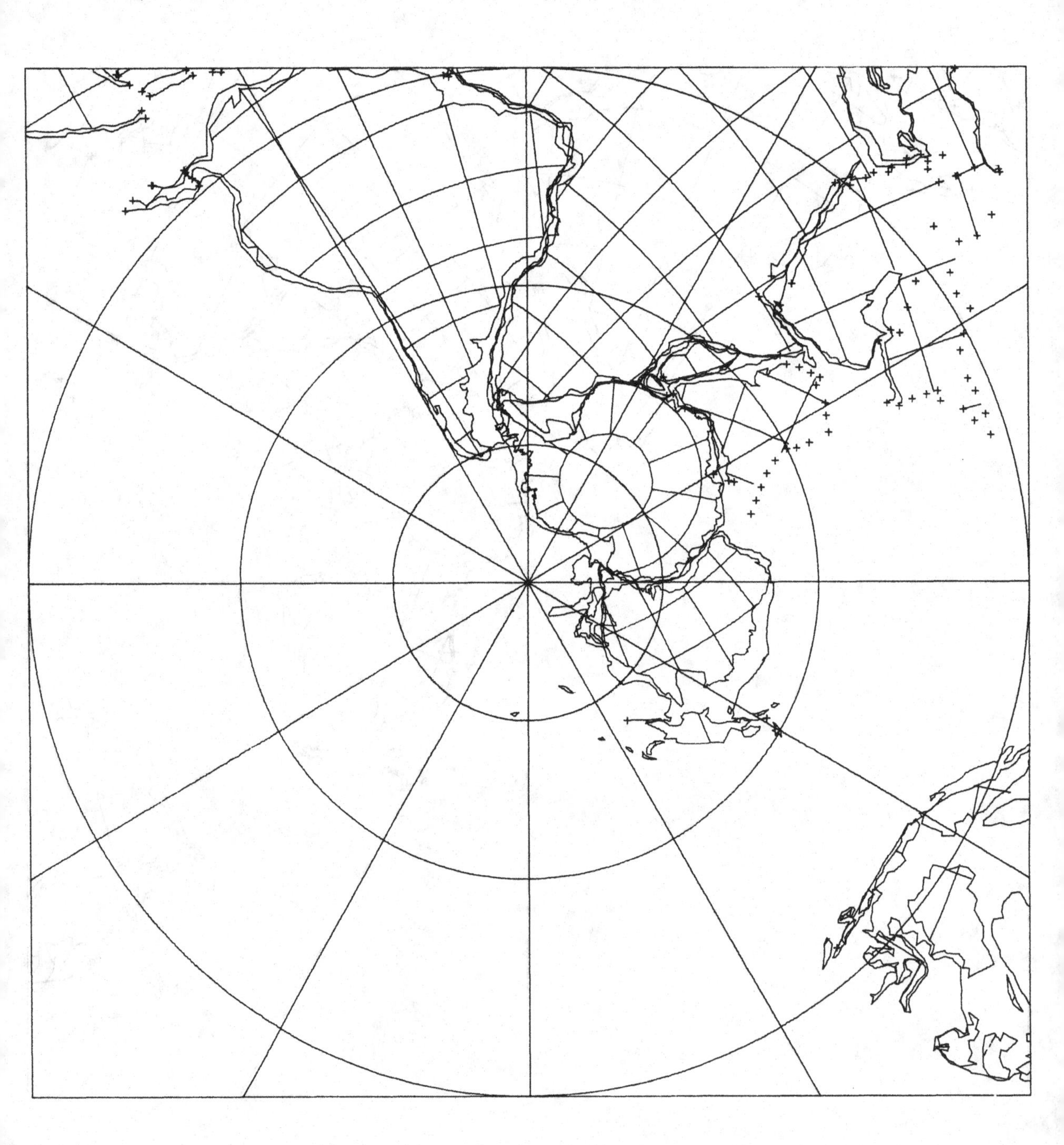

Map 36
160 million years
Callovian (mid-Jurassic)

South polar stereographic
$N = 18$ Alpha-95 = 8·7

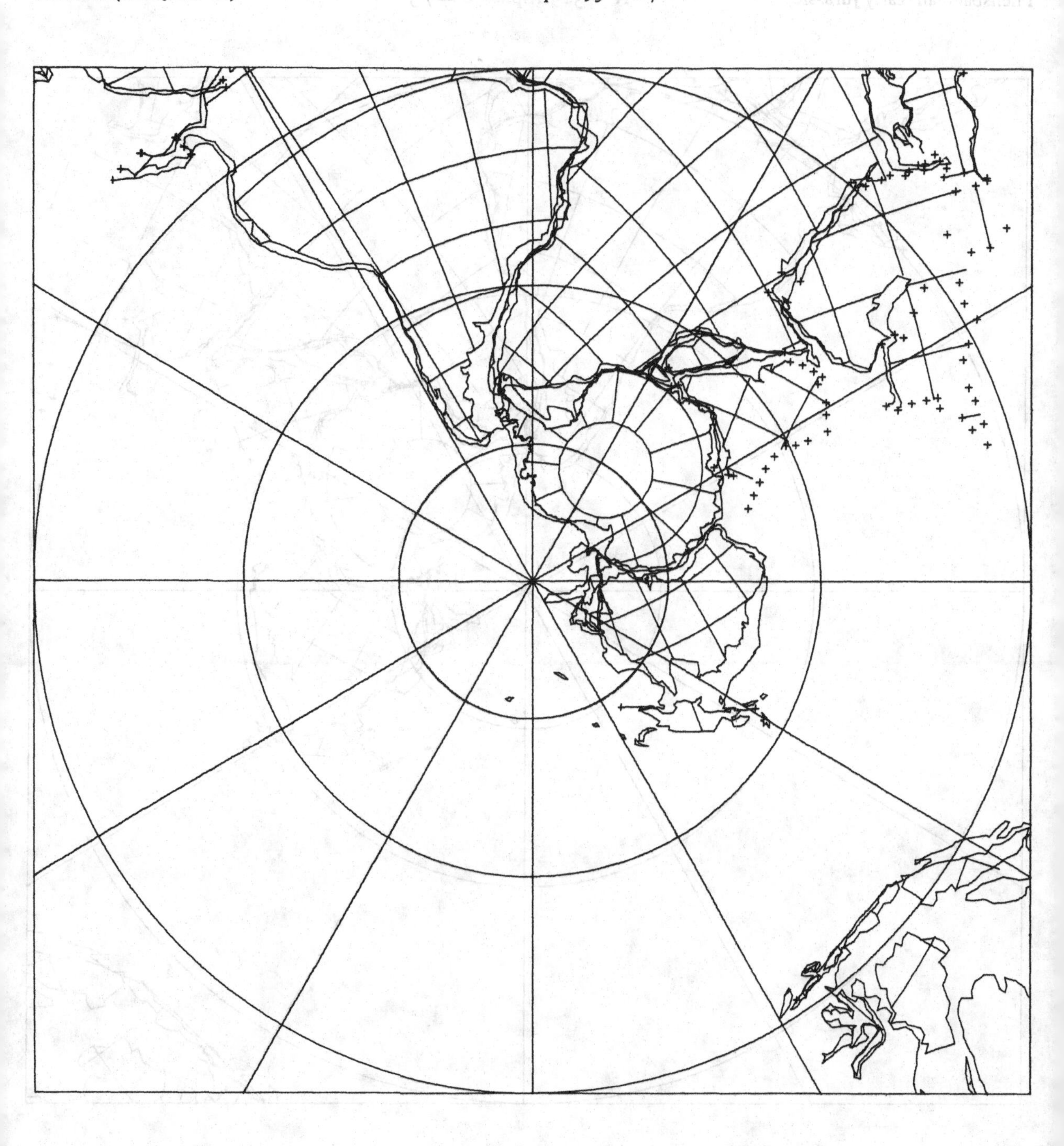

Map 37
180 million years
Pliensbachian (early Jurassic)

South polar stereographic
$N = 32$ Alpha-96 = 7·5

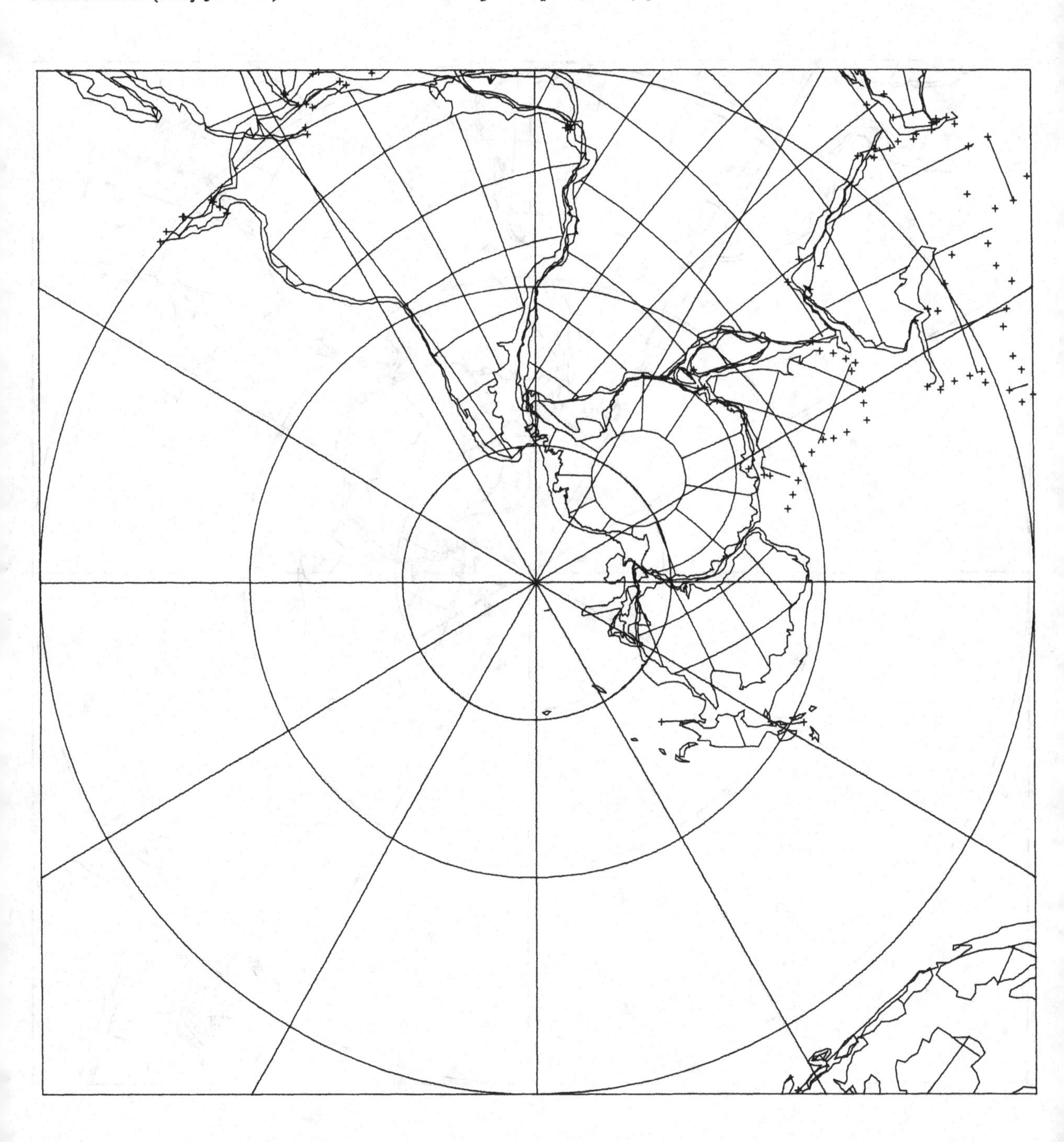

Map 38
200 million years
approx. Rhaetian (latest Triassic)

South polar stereographic
$N = 36$ Alpha-95 = 6·1

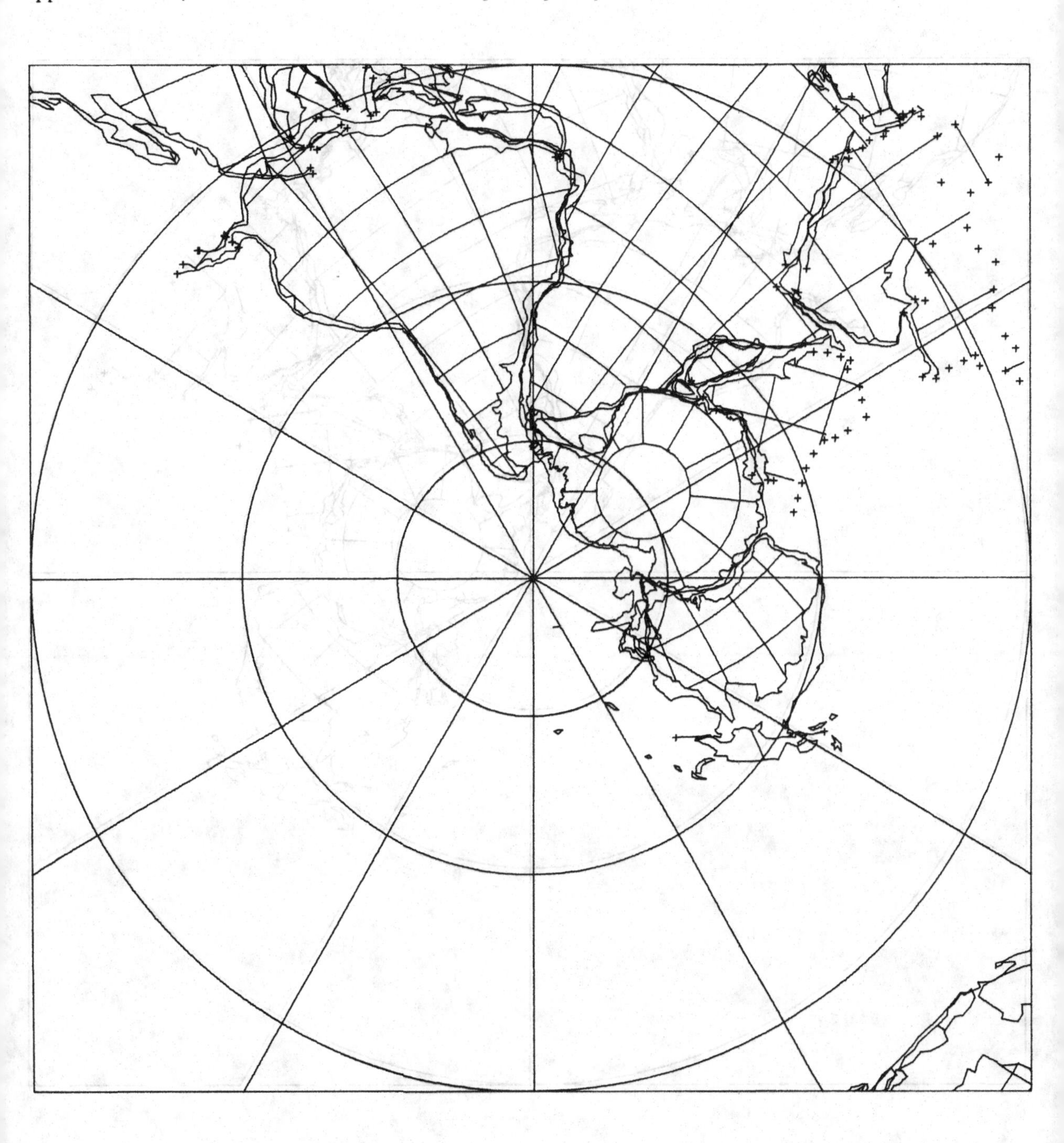

Map 39
220 million years
early Triassic

South polar stereographic
$N = 67$ Alpha-95 = 4·5

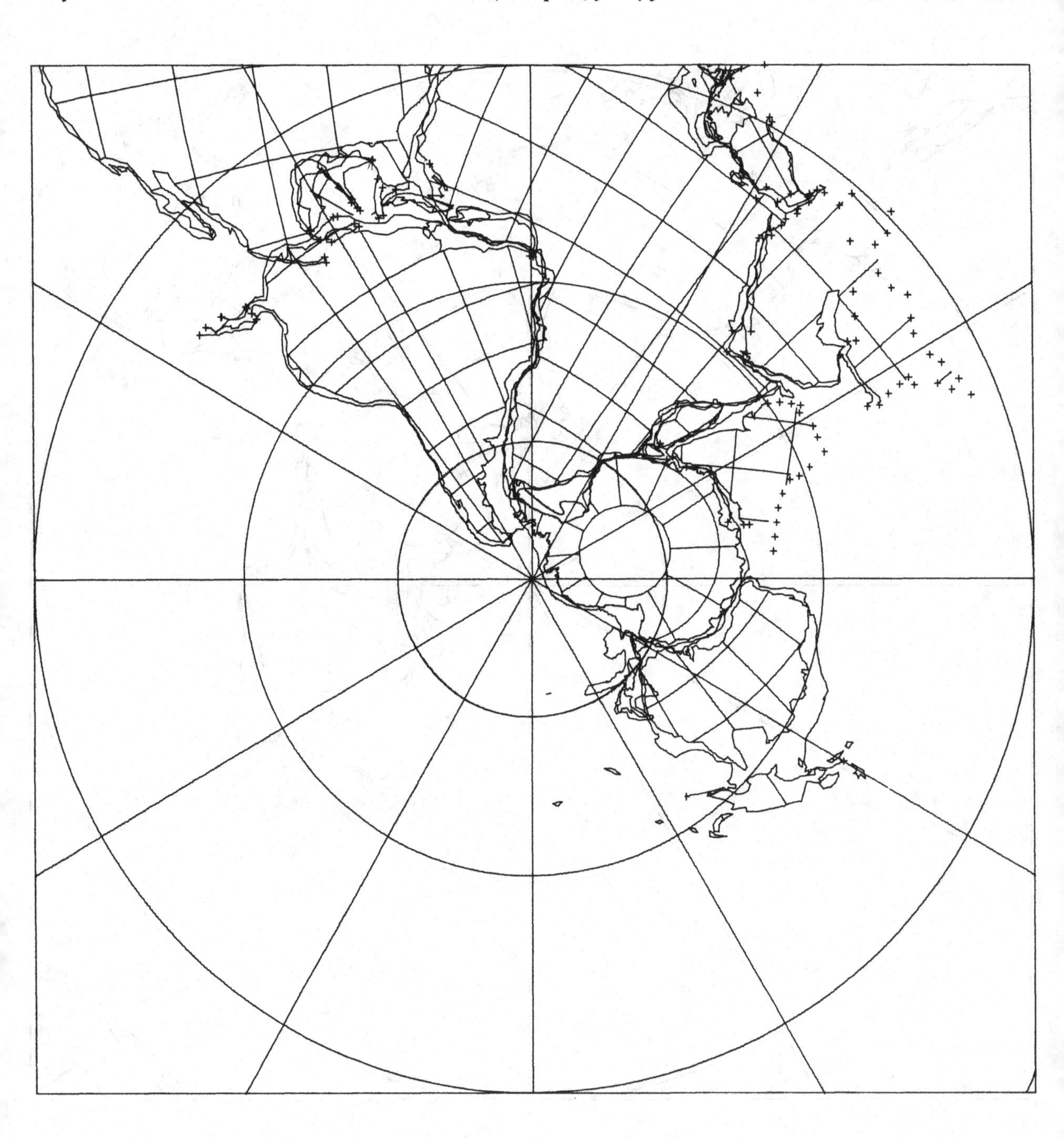

Map 40
Present day

Lambert equal-area

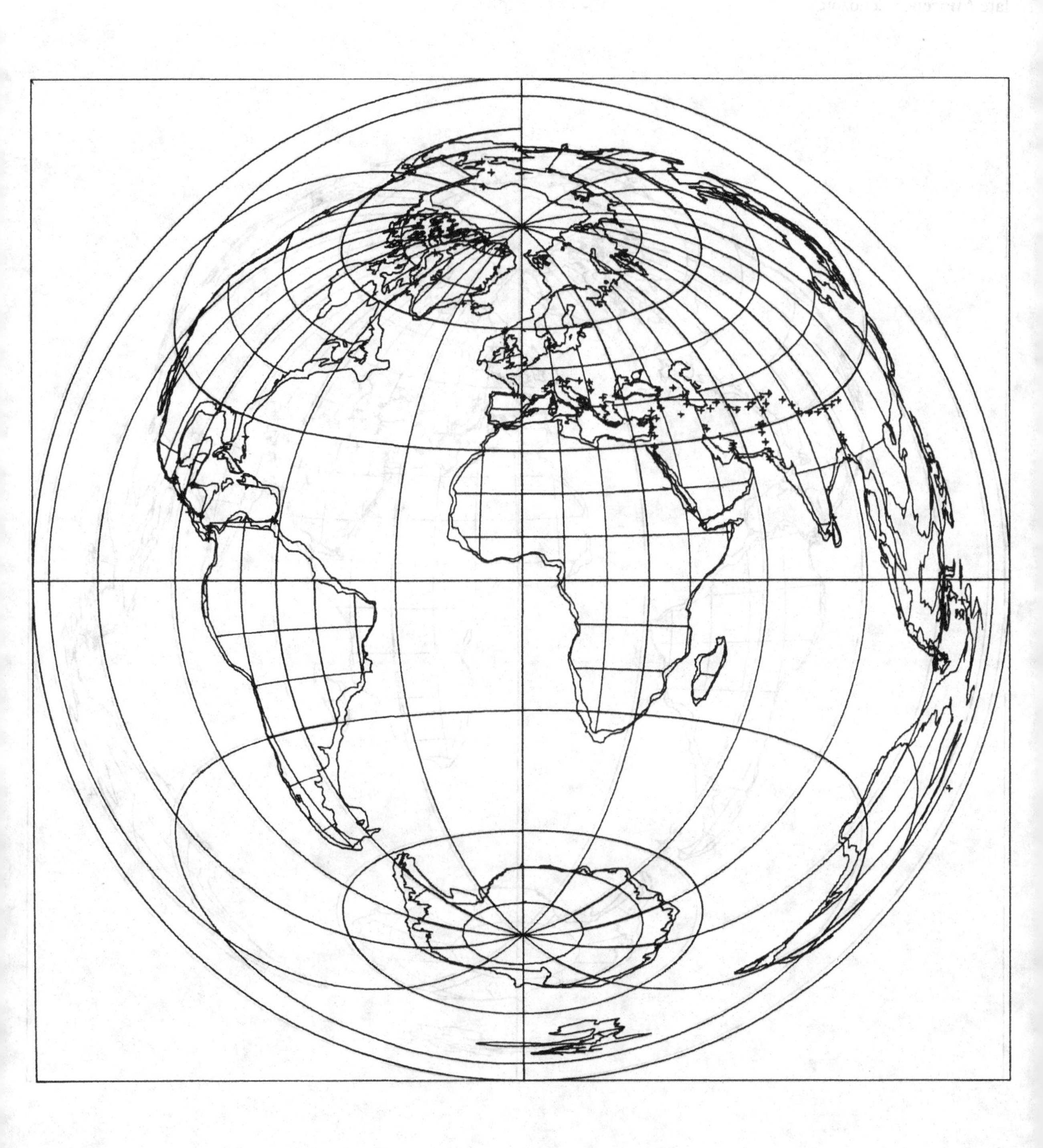

Map 41
10 million years
late Miocene (Cenozoic)

Lambert equal-area
$N = 87$ Alpha-95 = 3·2

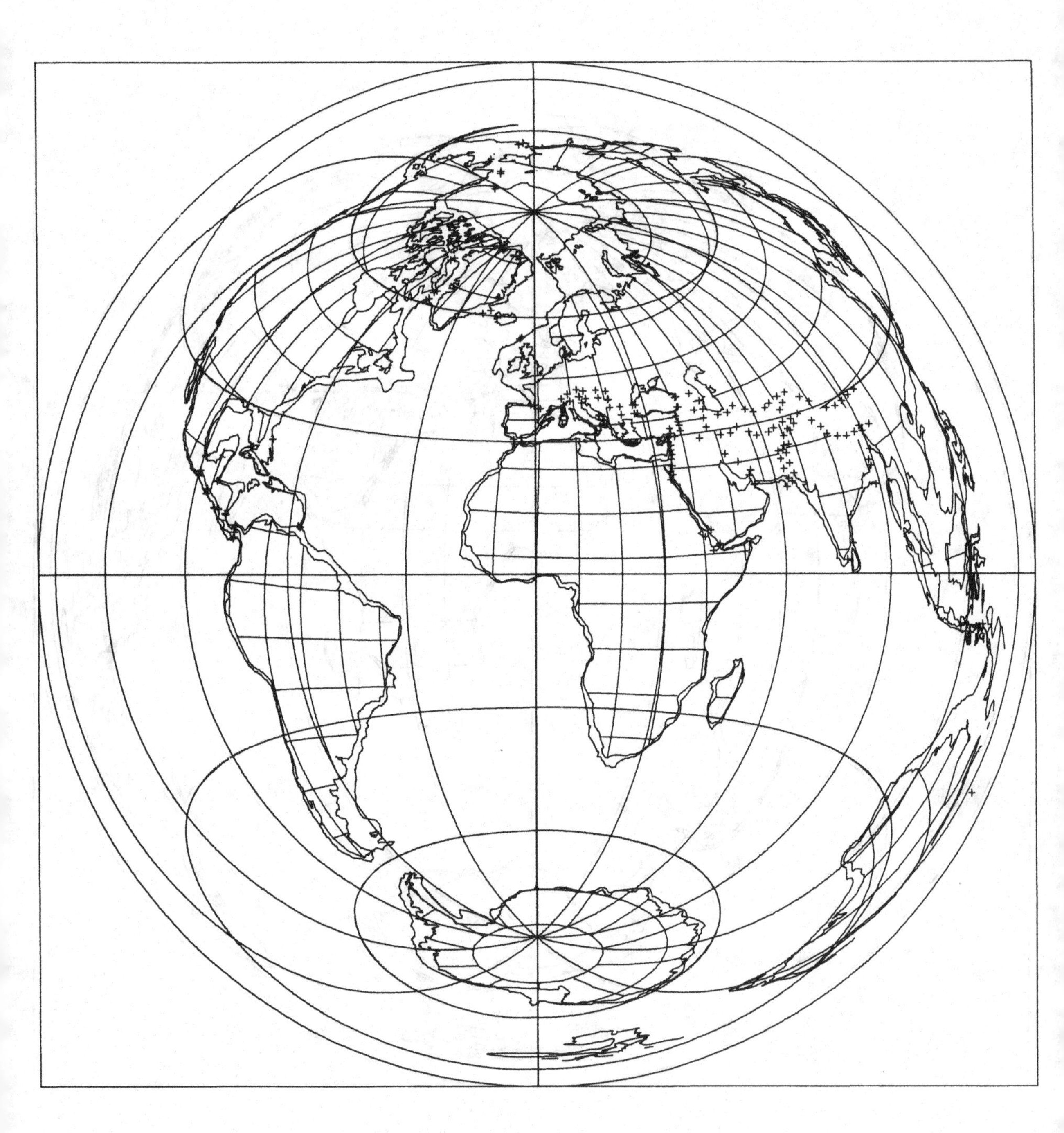

Map 42
20 million years
early Miocene (Cenozoic)

Lambert equal-area
$N = 46$ Alpha-95 = 5·7

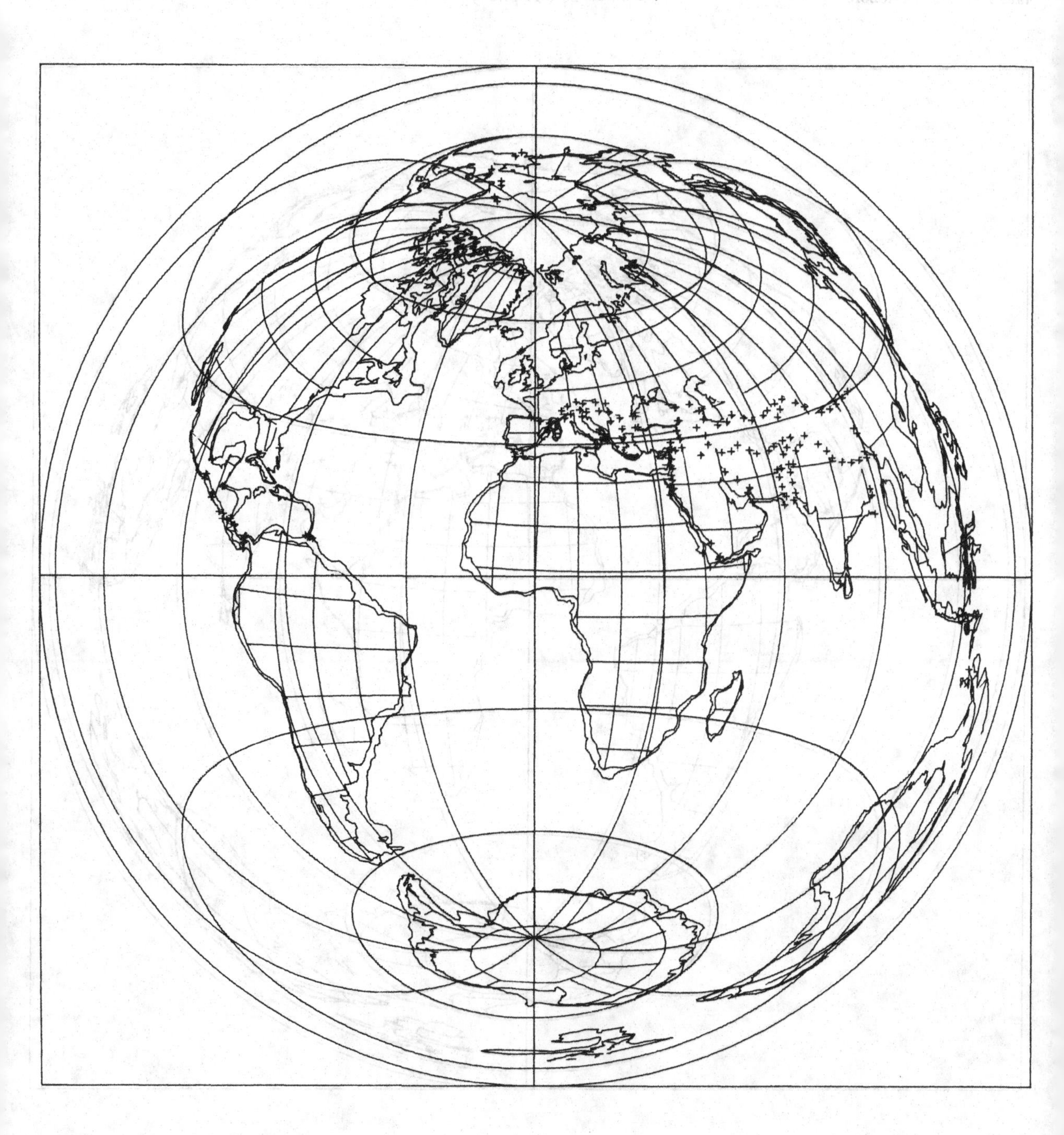

Map 43
40 million years
late Eocene (Cenozoic)

Lambert equal-area
$N = 28$ Alpha-95 = 4·5

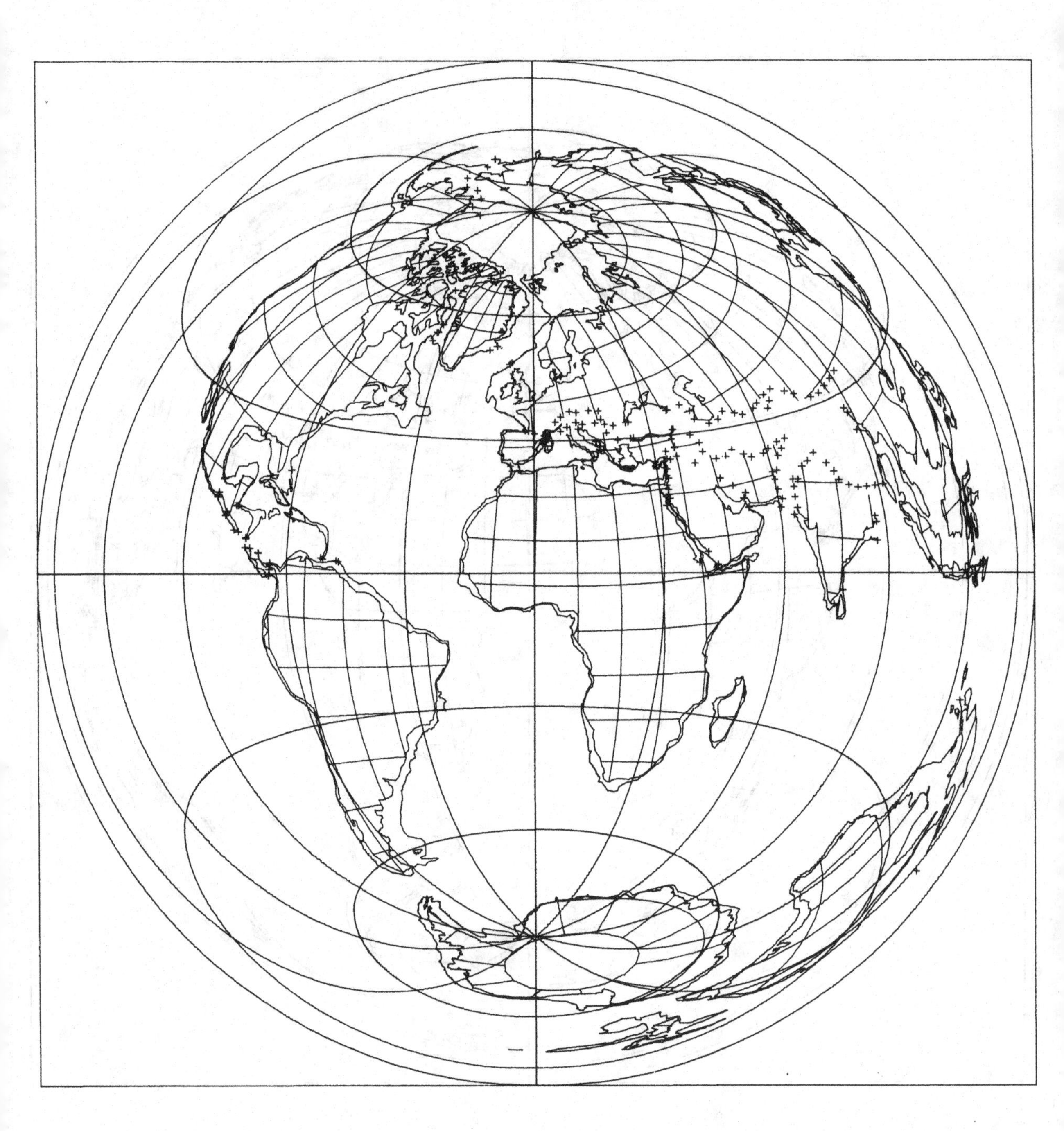

Map 44
60 million years
Paleocene (Cenozoic)

Lambert equal-area
$N = 43$ Alpha-95 = 4·7

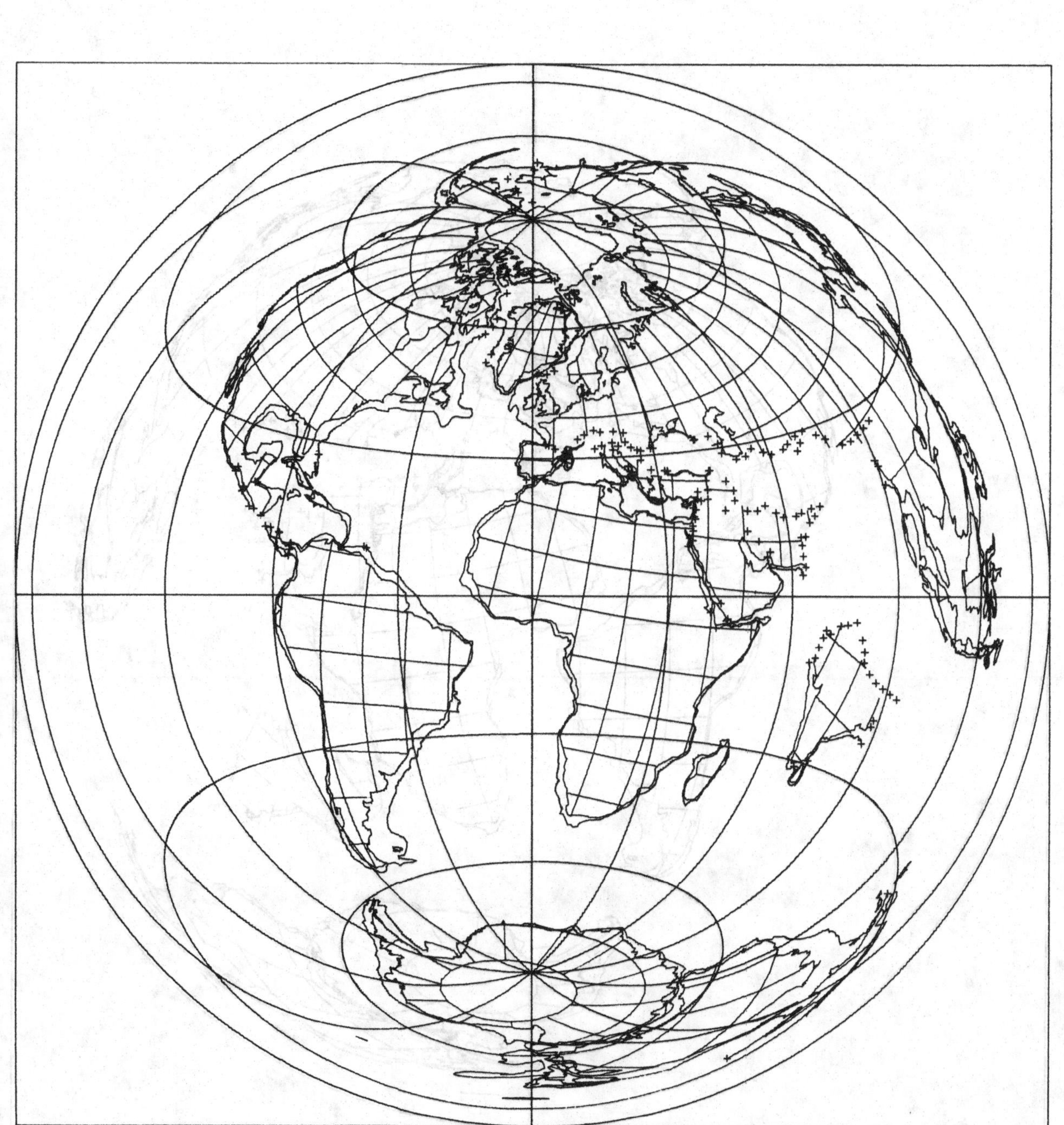

Map 45

80 million years

Santonian (late Cretaceous)

Lambert equal-area

$N = 25$ Alpha-95 = 6·0

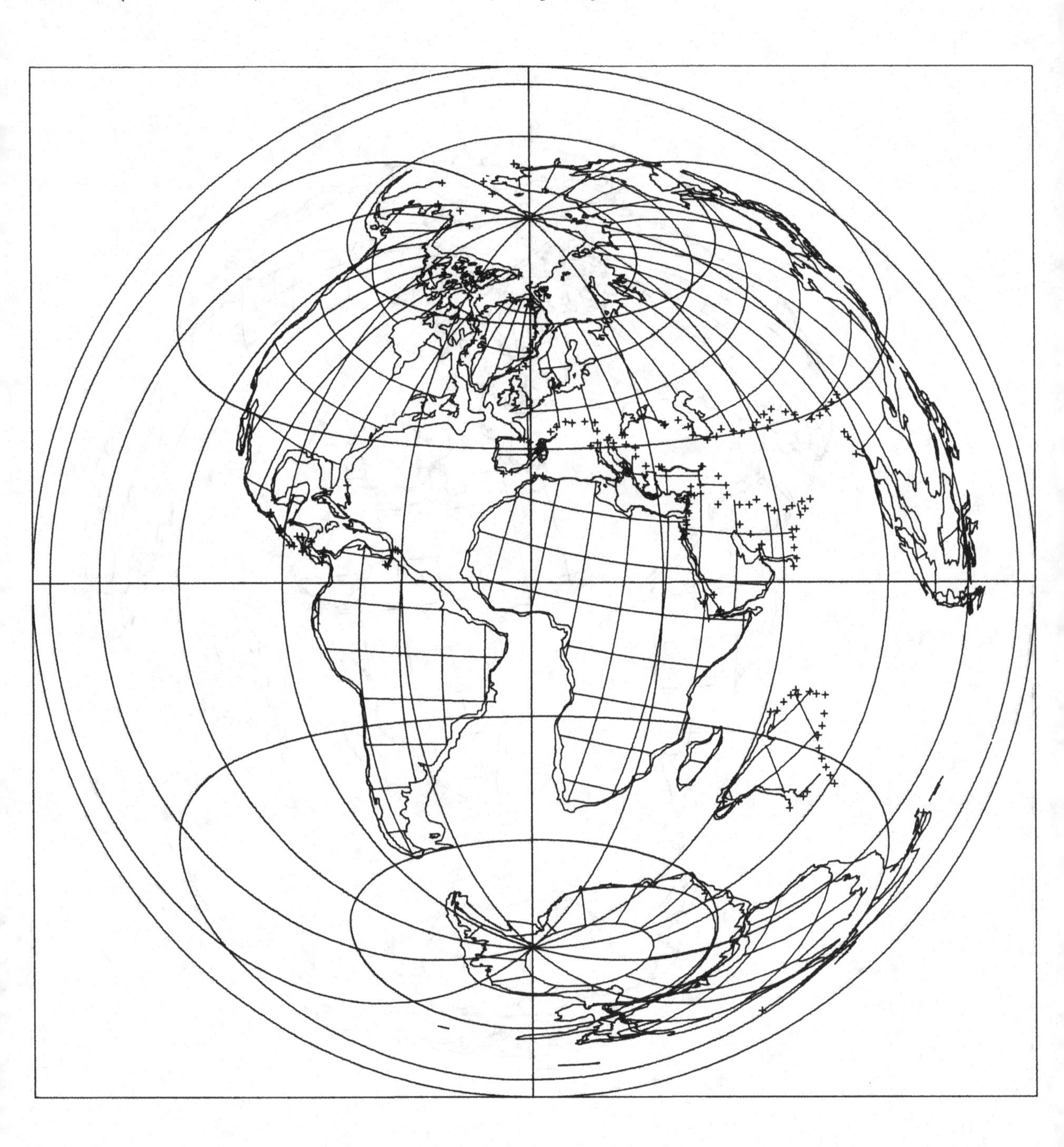

Map 46

100 million years
earliest Cenomanian (mid-Cretaceous)

Lambert equal-area
$N = 43$ Alpha-95 = 5·2

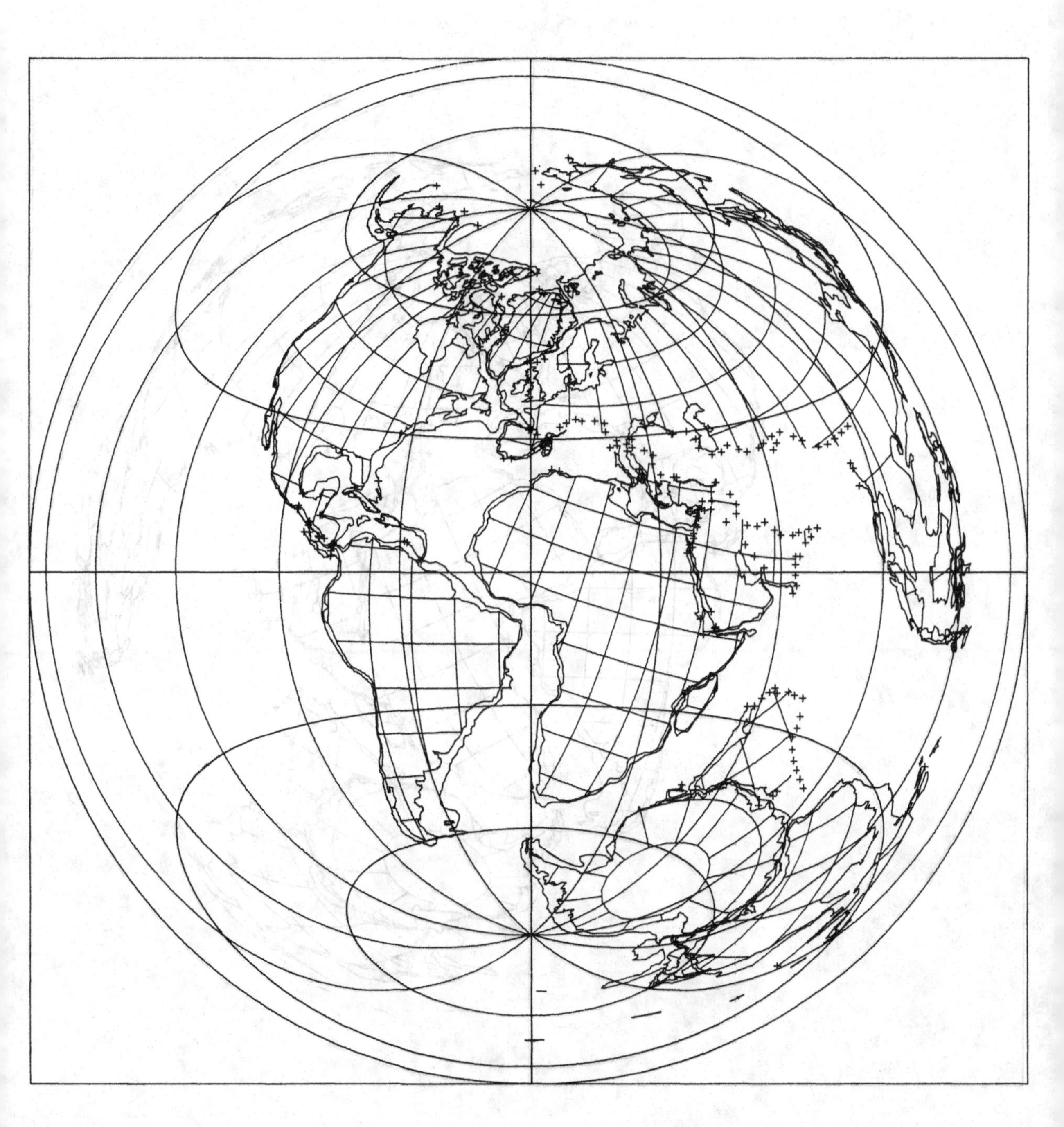

Map 47
120 million years
Hauterivian (early Cretaceous)

Lambert equal-area
$N = 27$ Alpha-95 = 6·2

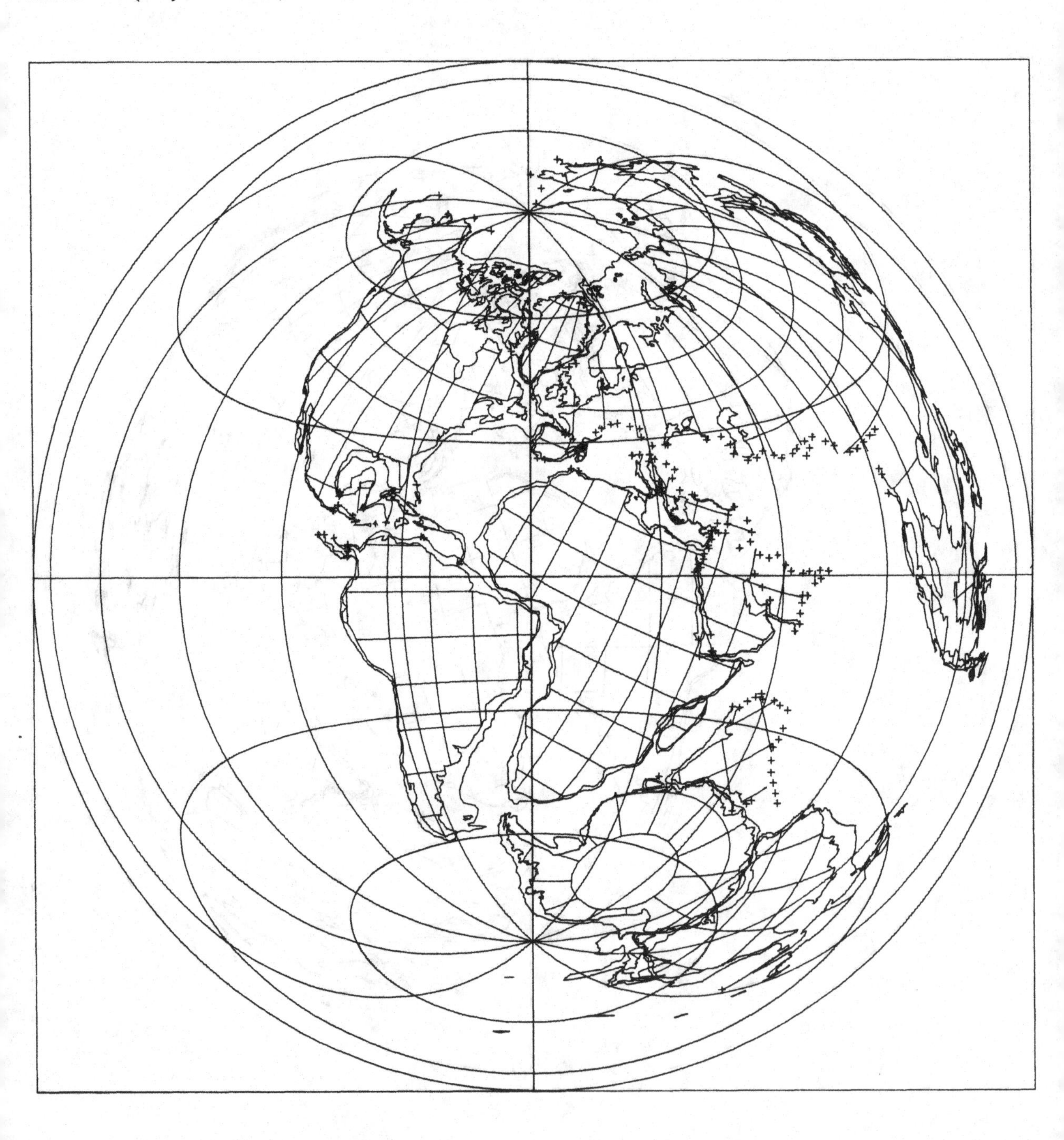

Map 48
140 million years
'Tithonian' (late Jurassic)

Lambert equal-area
$N = 33$ Alpha-95 = 5·4

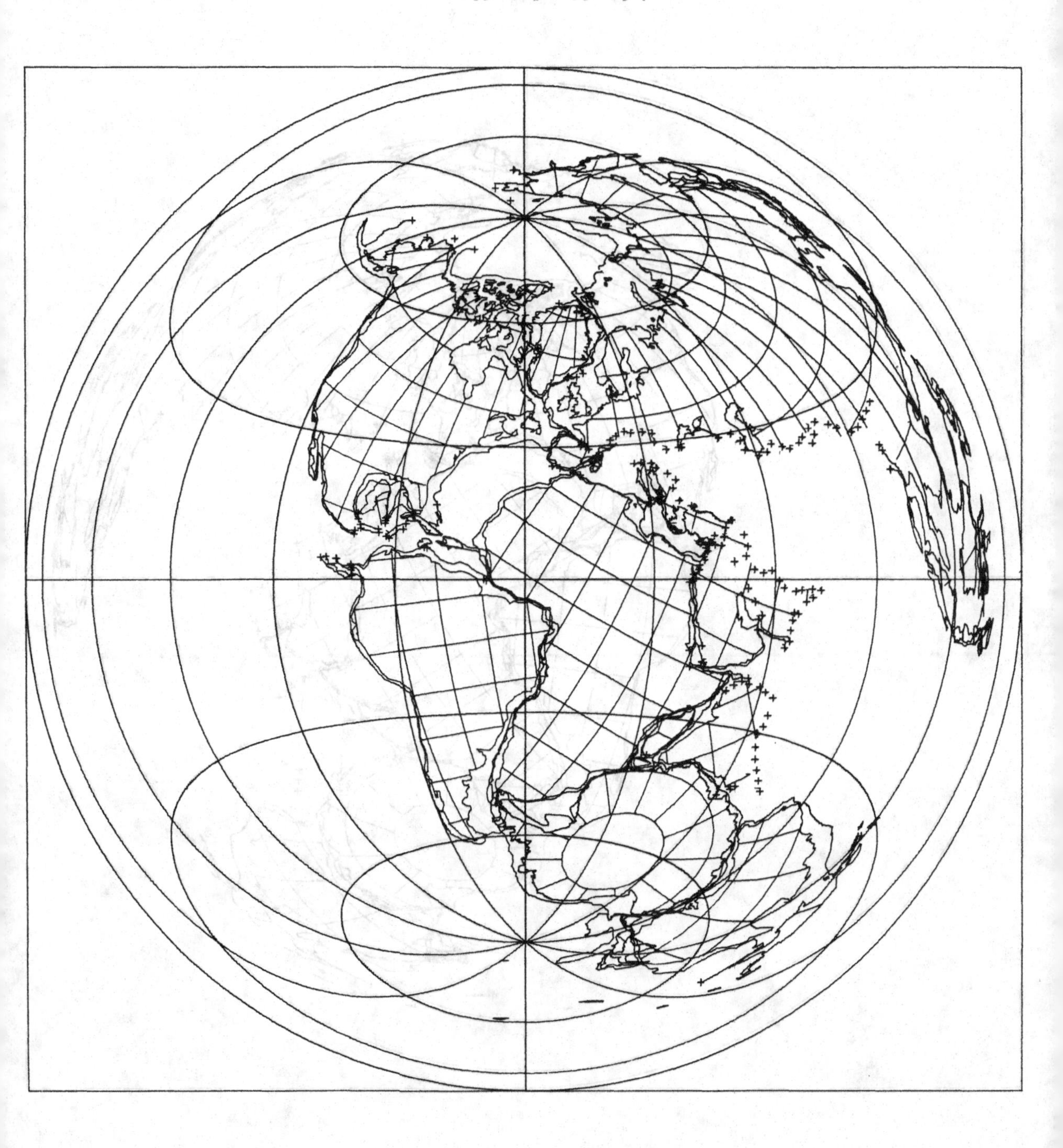

Map 49
160 million years
Callovian (mid-Jurassic)

Lambert equal-area
$N = 18$ Alpha-95 = 8·7

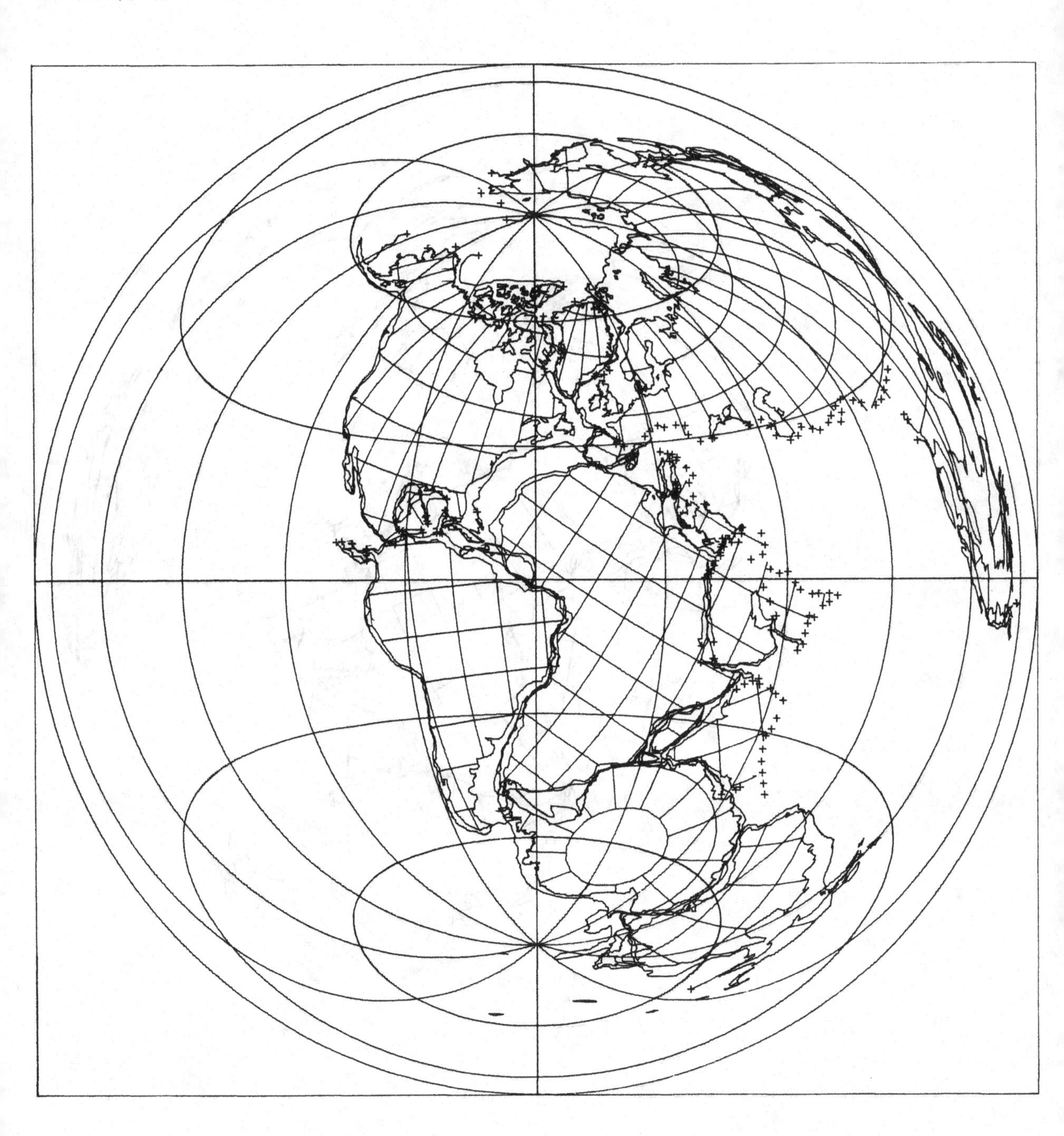

Map 50
180 million years
Pliensbachian (early Jurassic)

Lambert equal-area
$N = 32$ Alpha-95 = 7·5

Map 51

200 million years
approx. Rhaetian (latest Triassic)

Lambert equal-area
$N = 36$ Alpha-95 = 6·1

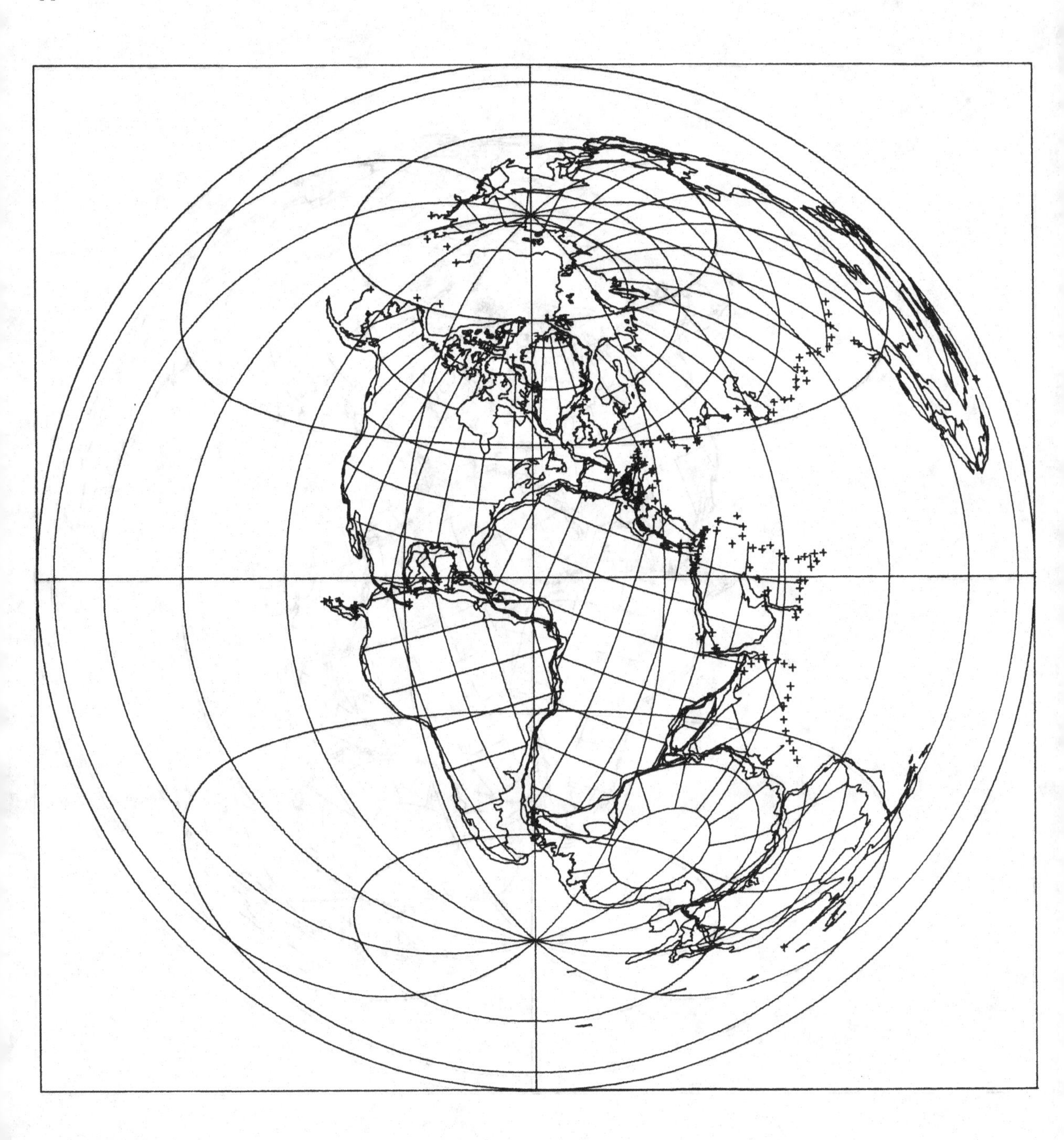

Map 52
220 million years
early Triassic

Lambert equal-area
$N = 67$ Alpha-95 = 4·5